ABOUT TURN,
FORWARD MARCH WITH EUROPE

IPPR/Rivers Oram Press

Already published

The Welfare of Citizens
Developing new social rights
Edited by Anna Coote

About Time
The revolution in work and family life
Patricia Hewitt

Strangers & Citizens
A positive approach to migrants and refugees
Edited by Sarah Spencer

Paying for Inequality
The economic cost of social injustice
Edited by Andrew Glyn and David Miliband

Educational Reform and its Consequences
Edited by Sally Tomlinson

ABOUT TURN, FORWARD MARCH WITH EUROPE

New Directions for Defence and Security Policy

Edited by Jane M.O. Sharp

IPPR/Rivers Oram Press

First published in 1996 by
Rivers Oram Press, 144 Hemingford Road, London N1 1DE

Published in the USA by
Paul and Company
Post Office Box 442, Concord, MA 01742

Set in 10/12 Sabon
by N-J Design Associates, Romsey, Hants
and printed in Great Britain by
T.J. Press (Padstow) Ltd

This edition copyright © IPPR 1996

The articles are copyright © 1996

Hugh Beach, Malcolm Chalmers, Michael Clarke, Andrew Cottey,
David Greenwood, Philip Gummett, Patricia Lewis, Neil Malcolm,
Colin McInnes, Stephen Pullinger, Jane M.O. Sharp, Ron Smith,
Trevor Taylor, Susan Willett

No part of this book may be produced in any form except for the
quotation of brief passages in criticism without the written permission
of the publishers

ISBN 1 85489 082 4
ISBN 1 85489 083 2 (pbk)

CONTENTS

LIST OF TABLES

ACKNOWLEDGMENTS

IPPR wishes to thank all who contributed to this book. An edited volume depends very much on the collegiality of the authors. The members of the 1995 study group that produced this volume for the Defence and Security Programme were generous with their time, and unusually cooperative in commenting on each other's chapters, in meeting deadlines, and in accepting editorial changes and cuts.

Special thanks also to Helena Scott, Publications Manager at IPPR, who shepherded the text through IPPR to Rivers Oram Press, to Katherine Bright-Holmes at Rivers Oram whose schedule was disrupted by the editor more than once, and to Gerald Holtham, Director of IPPR, who encouraged the project enthusiastically from inception to publication.

Jane M.O. Sharp, March 1996

LIST OF CONTRIBUTORS

The editor

Jane M.O. Sharp directs the Defence and Security programme at IPPR and is a Senior Research Fellow in the Centre for Defence Studies at King's College London. She was formerly a Senior Researcher at the Stockholm International Peace Research Institute (SIPRI) and National Director of the Council for a Livable World in Washington DC. Her publications include *Europe After an American Withdrawal* (Oxford University Press, 1990); *Bankrupt in the Balkans, British Policy in Bosnia*, IPPR, 1993 and *Conventional Forces in Europe (CFE): History, Analysis and Prospects for the CFE Treaty* (Oxford University Press, 1996).

The contributors

Sir Hugh Beach is a retired General (ex-Royal Engineers). He was formerly Master General of the Ordnance and Warden of St George's House, Windsor Castle. He is Vice-Chairman of the Council for Arms Control.

Malcolm Chalmers is a Senior Lecturer in the Department of Peace Studies, University of Bradford. He is the author of *Biting the Bullet: European Defence Options for Britain*, IPPR, 1993 and co-author, with Owen Greene of *Taking Stock: The UN Register after Two Years* (Westview, 1995). His current research interests include British defence policy, the UN register and security in South-east Asia.

Michael Clarke is Professor of Defence Studies and Executive Director of the Centre for Defence Studies at King's College, London. He was formerly a Senior Lecturer at the University of Newcastle-upon-Tyne. He has been a Guest Fellow at the Brookings Institution in Washington DC and an Associate Fellow at the Royal Institute of International Affairs. His recent publications include *British External Policy-making in the 1990s* (Macmillan, 1992) and *New Perspectives on Security* (Brassey's, 1993).

Andrew Cottey is a lecturer in the Department of Peace Studies at the University of Bradford. He has previously worked for Saferworld and for the British American Security Information Council. His research focuses on European security issues, the United Nations' role in international security and the concept and practice of conflict prevention. He is author of *The Pursuit of Peace: A Framework for International Action* (Saferworld, 1994) and *East-Central Europe after the Cold War* (Macmillan, 1995) and is currently working on a book on western security policy towards post-communist Europe.

David Greenwood is Director of the Centre for Defence Studies at the University of Aberdeen. His publications include *Budgeting for Defence* (1972).

Philip Gummett is Professor of Government and Technology policy at the University of Manchester. He has advised the House of Lords Select Committee on Science and Technology, as well as the European Commission, and the European Parliament, on defence technology affairs. His recent publications include, 'Nationalism, Internationalism and the European Defence Market' (with William Walker), Chaillot Papers No. 9, Western European Union Institute for Security Studies, 1993 and 'European Defence Technology in Transition: Issue for the UK', *Science Policy Support Group Review*, No.7, SPSG, 1993.

Patricia M. Lewis is Executive Director of the Verification Technology Information Centre (VERTIC). She was formerly a Lecturer in Physics at the University of Auckland, New Zealand. She has served as a consultant to the UN and to the Foreign and Commonwealth office. Her publications include 'Verification in a Labile World' in *Verification 1993: Peacekeeping, Arms Control and the Environment* (eds.) J. B. Poole, R. Guthrie, Brassey's, 1993; 'Organising for Effective Implementation', in *Implementing the Comprehensive Test Ban: New Aspects of Definition, Organisation* *and Verification* (ed.) Eric Arnett, SIPRI research paper 18, 1994; and 'The United Kingdom' in *Nuclear Weapons after the CTB: Implications for Modernisation and Proliferation* (ed.) Eric Arnett, SIPRI, 1995.

Neil Malcolm is Professor of Russian Politics and Head of the Russian and East European Research Centre at the University of Wolverhampton. From 1989 to 1993 he was Head of the Russian and CIS Programme at the Royal Institute of International Affairs. His most recent publications include *The New Eastern Europe: Western Responses*, with J. Rollo, J. Batt, B. Granville, 1990; *Russia and Europe: An End to Confrontation?* (Putnam, 1993) and *International Factors in Russian Foreign Policy*, with A. Pravda, R. Allison, M. Light, 1996.

Colin McInnes is Senior Lecturer in the Department of International Politics, University of Wales, Aberystwyth, and was formerly a lecturer in the Department of War Studies, the Royal Military Academy, Sandhurst. He is the author of *Trident: The Only Option?* and *NATO's Changing Strategic Agenda*. He has just completed a book on the British Army since 1945.

Stephen Pullinger is the Executive Director of the International Security Information Service (ISIS), an independent briefing service to parliamentarians on defence and security issues based in London and Brussels. He writes on nuclear and defence issues and was co-author of *Making the Commons Work*, IPPR, 1991.

Ron Smith is Professor of Applied Economics at Birkbeck College, London and has written extensively on defence economics.

Trevor Taylor is Professor of International Relations in the School of Social Sciences at Staffordshire University. He was formerly Chairman of the British International Studies Association, and Head of the International Security Programme at the Royal Institute of International Affairs. His publications include: *European Security and the Former Soviet Union: Dangers, Opportunities and Gambles*, RIIA, 1994; 'West European Security and Defence Co-operation: Maastricht and Beyond', *International Affairs*, Vol.70, No.1, January 1994; and editor of *Reshaping European Defence*, RIIA, 1994.

Susan Willett is a Senior Research Fellow at the Centre for Defence Studies, King's College London, where she is the Director of the Comparative Research Programme. Her recent publications include 'Controlling the Arms Trade: Supply and Demand Dynamics', *Faraday Discussion Paper 18*, CAC, 1991; 'Dragon's Fire and Tiger's Claws: Arms Trade and Production in far east Asia', *Contemporary Security Policy*, Vol.15, No.2, August 1994 and 'The Legacy of a Pariah State: South Africa's Arms Trade in the 1990s', *Review of African Political Economy*, No.64, 1995.

LIST OF ABBREVIATIONS

ABM	anti-ballistic missile
ACV	armoured combat vehicle
ARRC	Allied Command Europe Rapid Reaction Corps
ASW	anti-submarine warfare
BW	biological weapons
CBI	Confederation of British Industry
CFE	Conventional Forces in Europe
CFSP	Common Foreign and Security Policy
CIS	Commonwealth of Independent States (former Soviet Union)
CJTFs	Combined Joined Task Forces
COCOM	Coordinating Committees for Multilateral Export Controls
CPC	Conflict Prevention Centre (EU)
CSCE	Conference on Security and Cooperation in Europe
CSBMs	Confidence and Security Building Measures
CTBT	Comprehensive Test Ban Treaty
CVR(T)	Combat Vehicle Reconaissance (Tracked)
DCS	Defence Cost Study
DERA	Defence Evaluation and Research Agency
DESO	Defence Export Sales Organisation
DGA	Délégation Générale pour l'Armement
DRA	Defence Research Agency
DTI	Department of Trade and Industry (UK)
ECDG	Export Credit Guarantee Department
ECO	Export Control Organisation
EFA	European Fighter Aircraft
EPC	European Political Cooperation
EU	European Union
FCO	Foreign and Commonwealth Office (UK)

FSU	Former Soviet Union
FYROM	Former Yugoslav Republic of Macedonia
FYU	Former Yugoslavia
GARTEUR	Group for Aeronautical Research and Technology
ICBM	Intercontinental Ballistic Missile
IEPG	Independent European Programme Group
IGC	Intergovernmental Conference
IISS	International Institute of Strategic Studies
JCG	Joint Consultative Group
JCTF	Joint Command Task Forces
MD	Military District (former Soviet Union)
MIRV	multiple independently targeted re-entry vehicles
MLRS	Multiple Launched Rocket System
MoD	Ministry of Defence (UK)
MOU	Memoranda of Understanding
MT	military task
MUF	material unaccounted for
NACC	North Atlantic Cooperation Council
NATO	North Atlantic Treaty Organisation
NFU	no first use (nuclear weapons)
NPT	Non-Proliferation Treaty
NSA	Negative Security Assurance
NWS	Nuclear Weapons State
ORCI	Office in Europe for Research and Collection of Information
OSCE	Organisation for Security and Cooperation in Europe
PESC	Public Expenditure Survey Committee
PfP	Partnership for Peace
SSBN	Nuclear-fuelled Ballistic Missile Submarine
TLE	Treaty-limited Equipment
UNPROFOR	United Nations Protection Force
WEAG	Western European Armaments Group
WEU	Western European Union
WTO	Warsaw Treaty Organisation

INTRODUCTION AND SUMMARY

Jane M.O. Sharp

The next British government will face tough decisions with respect to its defence and security priorities. Recent cuts in the defence budget without commensurate cuts in military tasks have spread British resources dangerously thin. This volume explores which traditional roles and missions Britain can maintain, given its financial constraints, and which must be given up.

The Foreign Office likes to boast that Britain punches above its weight in the international system. With more commands in the North Atlantic Treaty Organisation (NATO) than the other European allies, and a permanent seat on the UN Security Council there is some truth in this. After the Second World War successive British governments tried to retain Great Power status by acquiring a semi-independent nuclear weapons capability and clinging to the apron strings of the United States. But the next British government must formulate policy on present realities rather than past glories and realise that it is now long past time when Britain should have set its course unambiguously for Europe. Relations with the United States will always be important for all the countries of western Europe, but Britain can no longer afford to invest its main effort in a special relationship which has not been special from the American side for many decades.

As (some) ministers in John Major's government are fond of saying, Britain should be at the heart of Europe. The alternative is not to be in a strong transatlantic relationship, it is to be a marginal offshore island of diminishing importance. The only way Britain is going to influence world events in the future is as a major European power working closely with France and Germany and dealing with the US as a power committed to Europe. Close cooperation will be especially important in the defence area. These three countries, with Italy and the Netherlands and others, have collaborated episodically on isolated projects, but if Europe is going to be able to cope with future European crises Britain must seek much deeper integration with France and Germany at all levels of defence planning and procurement.

An important task for the next government, of whatever stripe, must therefore be to engineer a better division of labour with our western allies. This means facing the difficult task of cutting not budgets, but traditional British defence roles. This will not be easy. New threats are replacing old ones and American reluctance to maintain its former commitments to NATO and the UN means there will be less support from the US than previously.

The old threat of a massive military attack from the Soviet-led Warsaw Pact has vanished, but new challenges are erupting not only from the turmoil in the former Communist countries, but also from resource shortages in the Third World which will create new challenges for the affluent north. Within NATO and the Western European Union (WEU), Britain and its allies must therefore work harder at developing a common set of principles and interests they are willing to defend, then decide who has comparative advantage in what and adapt national capabilities accordingly.

This book is organised around six themes. The first part looks at Britain's new challenges after the Cold War. The second section examines Britain's domestic priorities. The third looks specifically at the role Britain can play in Europe, including how Britain can contribute to the defence industrial base, develop a Common Foreign and Security Policy (CFSP) within the European Union (EU), and reassure the former Warsaw Pact allies who feel increasingly vulnerable to the turmoil in the Former Soviet Union (FSU) and in Former Yugoslavia (FYU). A fourth section examines Britain's role in the wider world, including relationships with the former Communist states and its potential contribution to preventive diplomacy as well as to more effective peacekeeping and peacemaking operations. A fifth part looks at specific arms control futures in which Britain has an important stake. Finally, Michael Clarke explores the issue none of the major parties has had the courage to address since the end of the Cold War: the costs and benefits of a non-nuclear Britain.

Part 1: Britain after the Cold War

Chapter 1 provides an overview of the likely challenges to British and European security in the twenty-first century. Michael Clarke lays out the threats and challenges to the UK posed by the new uncertainties of the multi-polar post-Cold War world relative to the stability and predictability of period from 1945—1990. He sees the UK under less direct military threat than almost any time in its history, but at the same time more deeply and intimately connected with the problems of European defence and security. Clarke sees six areas of potential unrest that could destabilise Europe: continued upheavals in the Balkans; the prospect of a Baltic crisis; conflicts in central Europe; a crisis in Ukraine, spillover from the continuing conflicts on the southern border of the Russian Federation and the prospect of unrest spreading north from the Mahgreb. Britain's interest in these potential trouble spots may be categorised as 'indirect', but 'deep'; a difficult circle for policy-makers to square.

Part 2: Domestic Priorities

Part 2 of the volume comprises four chapters that deal with domestic priorities. In Chapter 2 David Greenwood addresses the issue of roles, missions and resources. He asserts that John Major's government did not conduct a real defence review, but merely attached new labels to the military tasks assigned to British forces during the Cold War. Greenwood's chapter assesses the 50 tasks listed in the MoD Statement on the Defence Estimates (1994) and urges the next government to establish priorities and to rank roles and missions within the current budget constraints, not continue the policy of simply legitimising existing force structures.

In Chapter 3 Colin McInnes offers suggestions for restructuring the army, which more than any other service in Britain, faces the problem of matching new commitments to dwindling resources. McInnes warns that the army is in danger of becoming a hollow force. He sees the need to adapt to several changes, including the shedding of old commitments, like meeting the challenge from the Soviet-led Warsaw Treaty Organisation (WTO) and coping with terrorism in Ulster, but also taking on new ones like those already undertaken in the Gulf, in Former Yugoslavia and in Rwanda. McInnes would restructure the army away from a capability to fight high-intensity wars, which Britain in any event could not cope with, except in coalition with the United States, to focus on forces adequate for low-intensity and peace support operations.

In Chapter 4 Ron Smith examines some of the questions that

governments must address before the defence industry can provide the weapons and other military equipment deemed necessary to defend British interests. Smith concludes that Britain needs to devote more effort to cooperation with our European allies because national arms industries and national independence in procurement is no longer viable. A coherent defence industrial base is only feasible at the European level. Smith warns future governments not to set much store by grandiose plans to convert military plants to civilian production. Experience not only in Britain, but in many other countries, indicates that conversion does not work. It is as ridiculous to suggest that coal mines be converted to alternative production. Rather the need is to put the skills of the defence workers and managers to work in the civilian sector. Ron Smith questions whether arms exports represent an economic benefit for Britain considering the scale of the offsets, aid projects and other subsidies that must be offered to win sales. He proposes a Treasury investigation of the real national return on sales.

Susan Willett takes up the issue of British arms exports in greater detail in Chapter 5. Like Smith, she believes that weapons manufacturers' profits are largely at the expense of British taxpayers who finance the government subsidies deemed necessary to clinch sales. She describes how government policy falls far short of its own declared principles by selling arms to undemocratic governments with dubious human rights policies as well as to regions of instability where arms can only fuel conflict. Willett deplores the secrecy that surrounds British arms policy, which encourages abuse of declared guidelines. She recommends greater transparency and a greater responsibility for parliament to make ministers accountable. She notes that both the USA and Germany have more transparent and accountable export procedures than the UK. Recent arms sales scandals exposed by the Matrix Churchill affair and the Scott enquiry suggest that reforms are long overdue. These include an outright ban on production and sales of inhumane weapons such as anti-personnel mines, an audit of recent arms sales, more restrictive export controls more strictly applied, and a code of conduct among all the major exporters.

Part 3: Britain in Europe

Part 3 deals with Britain's role in Europe. In Chapter 6 Trevor Taylor outlines the effort to establish a Common Foreign and

Security Policy (CFSP) for the states of the growing European Union (EU). Beginning with European Political Cooperation (EPC) in the early 1970s, the European states saw the need to extend their cooperation in economic matters to wider political issues. EPC coordination proved especially important in the late 1970s and early 1980s with successful EC caucuses at the CSCE. The EU states have continued this tradition since the CSCE became the Organisation for Security and Cooperation in Europe (OSCE) in late 1994. An effective CFSP, however, requires more than episodic cooperation; it requires a common EU view of shared interests. EU policy towards Former Yugoslavia (FYU) suggests that there are substantial differences to overcome. Taylor urges the next British government to define the national interest more carefully taking into account both Britain's dependence on her European partners and the growing detachment of the United States from all its European allies, a theme that reverberates throughout this volume. Taylor believes the American connection is vital for Britain, but warns that the Washington policy-makers will not appreciate a parochial Britain, unwilling to engage with its EU partners in developing a coherent CFSP. To date both the Conservative government and the Labour opposition appear to favour intergovernmental policy-making with respect to CFSP and European defence. Unless EU states accept qualified majority voting, however, there is unlikely to be effective policy. Taylor suggests British policy towards former Communist Europe should be modelled on proactive Germany not cold-shoulder France.

Persuading Britain to make a more wholehearted contribution to European affairs does not necessarily mean spending more on defence. Indeed, in Chapter 7, Malcolm Chalmers argues that Britain carries a disproportionate financial defence burden compared to its European allies. With per capita income in the UK now below that of Germany, France and Italy, Chalmers urges a reduction of British defence spending to the same percentage of GNP as its main partners. There is a particularly strong message here for Germany to contribute more to the common defence, although Chalmers acknowledges that Germany contributes to international stability and security in other ways. Germany is a more generous donor of foreign aid, for example, and has taken in far more refugees from former Communist countries than Britain.

In Chapter 8 Philip Gummett echoes David Greenwood when he notes the British tendency to maintain all existing programmes, but at a uselessly stretched-out level. Gummett looks in more detail at Britain's potential contribution to a more coherent European-wide defence industrial base. Like Phil Smith, Gummett emphasises co-operation with Brussels rather than Washington. Gummett also urges the new British government to look hard at the Japanese model of defence production which turns out superior military equipment not by reliance on a specialised defence sector, but by maintaining a strong technology and industrial base.

In the last of the chapters on Britain in Europe, Jane Sharp looks at the contribution Britain can make to the assurances that western institutions, primarily NATO and the EU, can offer the newly independent former Communist countries in central Europe. Immediately after the end of the Cold War most of these states looked to the pan-European CSCE to provide for their security. After the CSCE's poor showing in the conflicts in the Former Soviet Union and Former Yugoslavia, however, the central Europeans began to look to western institutions. Despite the weak leadership of the Clinton administration, central Europeans particularly want to join NATO because they believe only the USA can provide the necessary bulwark against the challenges posed by both German and Russian power.

That said, the central Europeans were distressed by the lack of US leadership under the Clinton administration, which allowed Russia undue influence in the five-nation Contact Group seeking a settlement in Bosnia. Even when the US began to engage seriously in the search for peace in 1995, far from protecting the victims of ethnic cleansing, the plan under discussion in Dayton Ohio appeared more likely to codify the ethnic division of Bosnia as agreed in March 1991 by the main perpetrators of aggression, Serbian President Milosevic and Croatian President Tudjman. This is a far cry from the democratic pluralism that western states should be promoting in the former Communist countries.

Germany has been the western power that most consistently urges expansion of both NATO and the EU. France is lukewarm about offering support to the east, because it is preoccupied with what it perceives as threats from the south. Britain must obviously be sensitive to both German and French priorities, but its more immediate interest lies in supporting Germany's *Ostpolitik* to spread the benefits of the western security community eastward. This will also ensure that Germany does not rekindle historical memories by trying to settle central European problems on its own.

Part 4: Britain in the Wider World

As a former colonial power and a permanent member of the UN Security Council, Britain feels special responsibilities for security beyond the confines of Europe. Successive British governments have certainly not been shy of interfering in the affairs of other states around the world when British interests have been threatened.

In Chapter 10 Neil Malcolm argues that Britain is well qualified to take a pragmatic approach towards the former Soviet republics. He acknowledges that the unaccustomed restraint of Soviet behaviour under Mikhail Gorbachev makes the increasingly aggressive and authoritarian Russian behaviour under Boris Yeltsin especially hard for western Europe to cope with. Nevertheless he argues that there are many reasons why the western democracies should seek a cooperative relationship with Moscow. The problem has been that Russian government officials, media and intellectuals alike, even those who have been traditionally pro-western, tend to see a western hand in everything that has gone wrong in Russia since the collapse of the Soviet Union. Western governments certainly missed opportunities to help, notably in 1991 and early 1992 when Russia was in a more cooperative mood and again in early 1993 when President Clinton should have tried to engage Boris Yeltsin in a constructive settlement of the crisis in Former Yugoslavia.

Malcolm recognises the authoritarian, statist impulses that drive the Russian leadership, but nevertheless counsels patience in the west in dealing with a Russia humiliated by loss of empire, but still a nuclear superpower and the strongest single state on the Eurasian land mass. At the same time western leaders obviously cannot continue to condone Russia's repeated contraventions of international law and disregard for commitments undertaken as a successor state to the former Soviet Union.

In Chapter 11 Andrew Cottey urges the next British government to take the lead internationally in developing policies for preventive diplomacy. These range from fact-finding, mediation, human rights monitoring, political and economic sanctions and the preventive deployment of military forces. The costs of the UNPROFOR debacle—not only to the victims of aggression in FYU, but to the reputation of the UN, NATO and the EU, as well as the individual countries responsible—demonstrate that international efforts designed to prevent conflict are infinitely more cost-effective than efforts to resolve conflicts once they erupt. Indeed some would argue that once a conflict has broken out there is little the international community can do except mop up after a militarily decisive result. In some cases, and the UN effort on FYU appears to be one, the attempted cure may be worse than the disease.

British troops learned much from their experience with the UN Protection Force UNPROFOR in FYU. In Chapter 12 Hugh Beach looks at the new demands on British forces with respect to post-Cold War operations. Beach finds severely wanting the policy of the western security community towards a number of post-Cold War crises. As a former imperial power proud of its military expertise, Britain has so far been more adept at military interventions than preventive diplomacy. Prevention is obviously better than cure, but when aggressive rogue states grossly abuse human rights and change borders by force, intervention is often appropriate. Beach also suggests that the criterion of proportionality of response has little meaning when one side is conducting a genocidal war. He also notes that, contrary to the accepted doctrine of just war, intervention is more effective early than as a last resort. Beach suggests that Britain and France, as former imperial powers, have much to teach their western allies with respect to colonial policing and the effective use of minimum necessary force. He acknowledges, however, that confidence in United Nations peacekeeping has been undermined by the failed UNPROFOR operation and suggests that in future when peace enforcement is called for *ad hoc* 'coalitions of the willing' will be more effective than traditional UN peacekeeping operations.

Part 5: Adapting the Cold War Arms Control Regime

The current arms control regime consists of treaties and agreements concluded during the bipolar Cold War era when it made some sense to seek a balance of military forces between east and west. Britain has been a leader in the multilateral diplomacy that led to conclusion of a number of important international agreements, beginning with the Limited Test Ban Treaty in 1963, that were an important part of the detente relationship that developed between NATO and the Warsaw Treaty Organisation (WTO). In the post-Cold War era, however, some of these agreements look fragile, and need to be amended to meet new circumstances.

In Chapter 13 Jane Sharp reviews the status of the treaty on Conventional Forces in Europe (CFE) which sets equal limits on five categories of military equipment between NATO and the states that used to belong to the WTO. She makes the case for Britain to take the lead in persuading the other parties to revise the terms of the agreement to meet the legitimate security interests of all the European states in the post-Cold War era.

When the CFE treaty was signed in November 1990 it was seen in the west as a reflection of the new relationship between the western democracies and a newly conciliatory WTO under the benign leadership of Mikhail Gorbachev. The Russian military never liked CFE, however, and even before signature tried hard to undermine its provisions by moving equipment beyond the Urals out of the CFE zone of application. Under Gorbachev, military objections to CFE were kept in check, but Yeltsin allowed Russian military officers free rein to voice their objections. These have nevertheless been raised in the proper channels provided by the treaty and should have been addressed more sympathetically by the NATO powers. The treaty has been overtaken by events and should be revised not only because the Russians wish it, but also because CFE treaty limits are much more in western than in Russian interests.

Revising CFE would also be a practical way of making the expansion of NATO more palatable to Moscow, for example by not raising the NATO ceilings for treaty-limited equipment as NATO takes in new members. Rather, NATO's national ceilings should be adjusted downwards as the alliance expands so that NATO of the 16+ does not field any more heavy equipment than NATO of the 16.

In Chapter 14 Stephen Pullinger assesses British interests in controlling weapons of mass destruction. As a medium-sized power with fewer security problems than almost any state on the globe, yet retaining an independent nuclear arsenal, Britain has a special responsibility to consolidate the international nuclear arms control and nonproliferation regimes, as well as the regimes designed to control chemical and biological weapons. Pullinger notes British weaknesses and strengths in the international arms control effort. He argues for a more aggressive effort across the board, but especially to achieve a ban on nuclear testing. A strong British stand against further testing became even more important in 1995 as the new French president Jacques Chirac announced a resumption of French testing in the South Pacific. Japan, New Zealand and many other non-nuclear powers rightly condemned the French announcement as a betrayal of commitments—made by the nuclear weapons powers at the NPT review conference earlier in the year—to seek a comprehensive ban on testing.

Another area in which Britain, as a technologically advanced power, can take the lead is in the building of confidence that comes from reliable verification of compliance with international arms control agreements. In Chapter 15 Patricia Lewis outlines the principles and history of verification techniques and details a number of measures that Britain could promote in the interests of both national and international security. Lewis argues that Britain should be much more willing than previously to open up for inspection its nuclear weapons production facilities and weapons deployment platforms. Lewis also suggests that measures that have proved useful in verification can also be applied to preventive diplomacy.

Part 6: Rethinking the Unthinkable

Unlike Steve Pullinger, who assumes British nuclear weapons are here to stay, in the final chapter Michael Clarke questions the sacred cow of the independent British deterrent. He examines from first principles the rationale for continuing to maintain a British nuclear capability, assesses the political and economic costs and questions to need to continue with this particular defence task. Clarke recognises that nuclear policy has been particularly difficult for the Labour Party, but warns that the window of opportunity to move towards a far safer less nuclear reliant world will not remain open for ever. He sees the next few years as critical ones for Britain to set an example to the rest of the world to take nonproliferation seriously. If Britain continues to rationalise possession of nuclear weapons why should other less secure nations renounce them? Certainly in a European Union trying to develop a Common Foreign and Security Policy, eventually moving towards a common defence policy, two independent nuclear arsenals seems excessive.

PART 1

BRITAIN AFTER THE COLD WAR

1 SECURITY CHALLENGES

Michael Clarke

A new era in security policy

Since 1990 Britain has found itself in a completely different international security environment from the Cold War years. It can be characterised as one in which there are no longer any credible 'threats' to Britain, but all too many 'challenges'. Put simply, Britain no longer faces the threat of having to fight a war for national survival. In this respect the homeland is safer from attack, or from civil war aided by an external power, than at any time in the history of the British state. On the other hand, Britain's security interests are also more deeply and intimately connected with those of its European allies and partners than previously. This raises the question as to whether 'British security' implies a greater degree of involvement in other people's disputes and conflicts which in themselves are of only indirect relevance to British territorial security. Challenges to the security of Europe may not, of course, be most appropriately dealt with by military means. In an era that is both 'post-Cold War' and characterised by deep 'economic interdependence', the maintenance of international security requires the employment of many different economic, diplomatic and cultural techniques. In so far as military instruments remain relevant to British security, they do so in a quite new context. The need to be prepared to fight wars of survival has been replaced by the possibility of using military force as a matter of choice; employing force at our discretion whether it be in high-intensity wars, as in the Gulf in 1990–91, in civil conflicts, as presently over northern and southern Iraq, or in peacekeeping and 'peace support' operations, as in Cyprus, Former Yugoslavia, Angola, Rwanda, Mozambique, Cambodia, Western Sahara, Somalia, or Haiti.[1] As these examples demonstrate, if military force is used in the future, it will be at our discretion. These are the major premises on which future British security interests will be based.

To take the first premise, it is clear that Britain no longer faces the threat of a war for national survival. There is no plausible enemy on the European continent, still less a plausible motive for any hypothetical enemy to want to occupy the British homeland. In the modern world, territorial acquisition is no longer a source of wealth or a way of augmenting power; rather it invariably constitutes a drain on the resources of the conqueror and detracts from those economic sources of real political power which are the hallmark of the late twentieth century. Even a worst-case scenario, whereby an aggressive nationalist Russia emerges from the present chaos in that country, would not constitute a threat to the *survival* of Britain, as was generally the case during the Cold War.[2] Though it is possible to conceive (in such a worse case scenario) of Russia becoming involved in war in or around Europe, a Russian desire to conquer the British homeland is politically inconceivable and would be a military nonsense. Even in the case of a major European war involving, say, Russia fighting a number of other European states, Britain would still be facing a war of choice—where it had discretion on the level of its involvement—as opposed to a war for survival, such as was fought in 1939–45.

The only plausible threat to the British homeland for the coming generation is not of conquest, but of simple destruction of the homeland by the delivery of weapons of mass destruction through ballistic missile attack. This might be initiated either by a 'rogue state' in the less developed world, or perhaps a developed state under a temporary 'rogue leadership'. The technical capability to mount such a threat presently only exists in Russia and the USA. It will be at least another decade, at the very minimum, before other states—perhaps in the Middle East or North Africa—pose the same technological threat. The political intentions to activate such threats are easily overstated and, in truth, it is difficult to construct a sensible scenario in which a rogue state, be it an aberrant Russia or anyone else, could gain anything from a ballistic missile attack on Britain.[3] If exotic 'political blackmail' scenarios using ballistic missile attacks against the homeland *are* regarded as credible, then Britain and other West European states have at least a decade—probably longer—in which to develop appropriate policies to alleviate or remove such threats.

A number of such policies to deal with the ballistic missile threat suggest themselves, not all of which are compatible; some of which are mutually exclusive. They include: the continuation of independent nuclear deterrence, a more active nonproliferation policy, a collective political approach to cope with rogue actors, a broad front political and economic approach to strengthen international regimes which would inhibit both the possession and use of such technologies, or the construction of a ballistic missile defence system covering either parts of Europe or perhaps merely parts of the British homeland. Since these policies cannot all be successfully pursued simultaneously, clear choices would have to be made in order to address the problem. But ballistic missile scenarios should not be regarded as an immediate, likely or grave threat to British security. At worst, there is the potential to develop into a threat at some time in the medium-term future.[4]

Of course, all circumstances can change, and the riposte of the cynic to such relatively sanguine analyses may be that war was not expected in the first decade of the twentieth century, or in the 1930s, or prior to 1982, or indeed in 1990. But the cynical view must take account of two fundamental facts. First, though wars may take place in which Britain is involved, they will not, on present trends, be wars for national survival, which states have little choice but to fight. Secondly, though all historical circumstances can and must change, defence policy has to make realistic planning assumptions. The argument here is that our present—and foreseeable—security environment offers no evidence for an assumption that we must plan to fight a war for national survival. The onus is therefore on the cynics to demonstrate that the circumstances of the emerging twenty-first century are altering to such a degree that we should amend the assumption. The mere assertion that 'history is full of surprises' is not good enough.

The second premise, however, represents the other side of the equation. It is that in the post-Cold War world there are many more challenges to British security *interests* in a world environment in which there are over 30 ongoing wars and in a Europe in which there are around 60 potential conflict points arising out of disputed territories, significant ethnic or national minorities, or active threats of civil strife. Three-quarters of such potential conflicts lie east of the old Iron Curtain.[5]

It is a reasonable assumption that the most important challenges to British security interests lie in Europe. It is conceivable that Britain may face a challenge to its security interests on a more global basis, somewhere else in the world: perhaps again in the

Falklands, in East Asia, or in the service of the United Nations, wherever the UN may become further involved in world conflicts. British economic, diplomatic, and perhaps military activity, will certainly continue throughout the world in the coming decades, creating interests which the state may wish to defend. But even if the promotion or protection of such activities is regarded as a security interest proper, it is nevertheless the case that they will constitute security interests in which there is a high degree of discretion. Britain may choose the degree to which it identifies its own security interests with those of the burgeoning role of the United Nations; or it may choose the degree to which the promotion or protection of economic interests are important matters of national security. The evacuation of British nationals from dangerous situations around the world may be regarded as a perennial security interest which exists on a global scale, but even this is peripheral—certainly not in itself a reason to maintain extensive military forces—and the protection of nationals abroad can be pursued in a number of different ways. In short, the security problems which matter to Britain and which create, one way or another, inescapable problems are those in and around the continent of Europe.

Post-Cold War Europe and its periphery displays both stabilising and destabilising tendencies. If some of the political landscape of the nineteenth and early twentieth centuries is beginning to re-emerge in Europe in the form of nationalism and ethnic fragmentation, it should also be noted that the continent simultaneously displays some unique historical trends which cut across the nationalist landscape. The late twentieth and early twenty-first centuries constitute an era of enormous economic interdependence which is at its most intense form in modern western and central Europe. Economic interdependence does not necessarily stop the occurrence of wars, since wars reflect the extremes of social hostility, but it increases the price of war both to participants and neighbours, and severely diminishes the economic benefits of mere territorial conquest. In developed and economically sophisticated societies, economic interdependence is certainly a disincentive to outright war. Then, too, Europe has never before in its history been so institutionalised. No other area of the world exhibits such a concentration of international institutions, multinational investment, and significant nongovernmental organisations. Though the individual states of Europe in both east and west are still major players in European politics—they provide the permissive framework within which other political activities can or cannot take place—no other group of states in the world have to take so close an account of the institutional and economic structures which condition their policy choices. The fifteen existing states of the European Union represent—with minor exceptions—a core of peaceful and prosperous European states who have formed a 'security community' wherein the prospects of war between them is simply unthinkable for the foreseeable future. Some nineteen other European states (excluding Russia) desire to join this western economic and security community in some way by the early part of the next century to share in the benefits. Whether or not this is a feasible proposition, the extent of the desire to join the western community suggests that NATO and European Union governments have at their disposal a powerful incentive to encourage peaceful behaviour.

This encouraging long-term picture, however, is small comfort in the shorter term, since economic interdependence, international institutionalism, and the incentive to join prosperous security communities are difficult to manipulate for the good in short-term crises. More to the point, such stabilising factors are far less effective on the leaderships of subnational groups as opposed to states themselves. Many of the present instabilities in Europe are triggered not by governments but by subnational groups, and some governments are simply too politically weak to restrain violent non-state forces. When western governments can only offer long-term carrots and short-term sticks (rather than vice versa) their range of policy options in dealing with immediate crises of insurgent nationalism, ethnic fragmentation, or even territorial disputes resurrected from a distant past, are somewhat limited. International economic sanctions against Serbia, for example, had a discernible effect on the attitude of the government of President Milosevic in Belgrade towards the Bosnian crisis, but very little effect on Mr Karadzic or General Mladic leading the subnational group of Bosnian Serbs fighting within Bosnia itself.

The potentially destabilising trends in Europe over the next three to eight years appear to fall into six main areas.[6] First, there is the distinct possibility of further conflict in the Balkans regardless of the eventual outcome of the war in Bosnia. The fate of Kosovo is critical, containing as it does some 90 per cent Albanian Muslims who are, in effect, oppressed by a 10 per cent Serbian population in a territory which the Serbian government regards as geopolitically and culturally vital to its survival. The Former Yugoslav Republic of Macedonia (FYROM) also represents a potential conflict with its ethnic mix of populations and the competing historical claims of its neighbouring states. It is difficult to imagine a conflict either in Macedonia or Kosovo being insulated from the other, and unlike the Bosnian conflict it would prove difficult to prevent neighbouring states from becoming overtly involved in any fighting. What Huntington has described as wars arising out of the 'kin-country syndrome' seem all too likely in this region.[7] The possibility of a Balkan conflict—if not a war—involving a number of Balkan states, two of whom are NATO members, is therefore still a possibility.[8] To this general scenario must be added the increased tension in recent years between Greece and Turkey over bilateral issues in the Aegean and in respect of the Turko-Kurdish civil war. Though a Greco-Turkish war over specific bilateral issues appears unlikely, heightened hostility between them may make involvement in other Balkan crises both more possible and more widespread.

Second, there is the prospect of a Baltic crisis, particularly between Russia and Estonia, and perhaps Russia and Latvia, over the position of the Russian minorities in those countries (around 40 per cent and 32 per cent respectively). A direct Russian invasion of the three Baltic states, or an official attempt to incorporate them into the Russian Federation is unlikely, but a breakdown of order in one or more of the three states, in which significant Russian help, both public and private, was offered to the Russian minorities in Estonia and Latvia constitutes a plausible crisis scenario which would raise many difficult questions for European and particularly Scandinavian security interests.[9]

Third, there is the prospect of conflict in central Europe, occasioned (though not necessarily created by) the 3.5 million Hungarians living as minorities in at least seven other European states. In particular, 2 million Hungarians live in the Transylvanian Region of Romania, some 600,000 in southern Slovakia, 450,000 in Serbia, and 160,000 in Ukraine. Other smaller Hungarian minorities exist in Austria, Slovenia, and Croatia. The Hungarian minorities in Romania (which has a population of around 22 million) and southern Slovakia (where the population of Slovakia is 5 million at most) are particularly significant, since in both cases they exist in definable geographical areas, where a high level of cultural autonomy may reasonably be demanded, and elements of political autonomy are repeatedly mooted by the Hungarian government.

Fourth, a crisis of major proportions is a possibility in Ukraine, either through an economic and social collapse in that country, or else in a Russian-Ukrainian crisis, which could not be resolved and which occasioned an effective collapse or dismemberment of that country. There are many existing sources of tension between Russia and Ukraine, from the implementation of the division of the former Soviet Black Sea fleet, the implementation of the tripartite nuclear accord which should denuclearise Ukraine, and more significantly over the future of the Crimea with its predominantly Russian population. So far, Russia and Ukraine have resolved—at least in outline—some of the most immediate differences between them, but the more difficult issues remain, underlying all of which is a deep-seated Ukrainian scepticism that Russia genuinely accepts the sovereignty of the new Ukrainian state. Even without such pressure, however, Ukrainian unity cannot be guaranteed, since there are politically significant Polish, Bulgarian and Tartar minorities in addition to the critical Russian minority in Ukraine, claims against Ukrainian territory from Moldova and a self-proclaimed Dniestr Republic which is a Russian-dominated enclave in Moldova containing a majority of Ukrainians. If Ukraine's long-term economic prospects are reasonably bright, it is nevertheless highly dependent in the short-term on Russian energy and infrastructure and has not received either the degree or type of western economic aid which would make a significant difference to its short term economic prospects. The importance for central Europe of a Ukrainian crisis—whatever form it could take—lies in the instability it would create for the Visegrad states of Poland, Hungary, the Czech Republic and Slovakia. The pressure of such a crisis on their eastern borders would not only be disturbing in itself, but would likely encourage further ethnic fragmentation of central European minorities who would take note of the fate of the various ethnic minorities in and around Ukraine itself. The collapse or severe disruption of any European state as large as Ukraine, having as it does a population of some 60 million, which puts it on a par with the Big Four in western Europe, would therefore constitute a significant crisis throughout central Europe which could have major effects on some of the states of western and southern Europe.

Fifth, there is the prospect of continuing conflict along the southern borders of the Russian Federation, in particular throughout Transcaucasia, which may involve inter-state crises in the region and will almost certainly have an impact on Russian policy. Russian forces are already fighting ongoing civil wars in Chechenya (within the Russian Federation) and in Tajikistan on behalf of the Tajik government, having already made a decisive intervention in the civil conflict in Georgia. A nationalist Russia suffering increasing pressures towards fragmentation from within and extensive civil conflict on its borders may become a very awkward partner for other European states. Given that Russia has so many common borders throughout Europe and is becoming increasingly concerned that it could be surrounded by adversaries in Europe, there is a prospect that instability and fragmentation on one of its borders could provoke a nationalist response on another. A Transcaucasian crisis, in other words, could act as a trigger for other crises involving Russian minorities elsewhere. Further crises in the Transcaucasus will, in any event, have a likely effect on Armenia, Azerbaijan, Turkey and Iran.[10]

A sixth potential security challenge is the growing immigration pressure on Europe both from the east and the Mediterranean basin in which large numbers of economic and political migrants encourage security clampdowns at significant borders and may fuel domestic political backlashes within Europe. Though attention has been given since 1990 to the prospects of westward migration from countries of the former Warsaw Pact and the Soviet Union, the more significant immigration pressures arise in North Africa. Certainly, the Mediterranean states of western Europe—Spain, France, Italy and Greece—are becoming increasingly preoccupied with Mediterranean security problems as opposed to those arising directly from central and eastern Europe. In this respect, western Europe may be shifting from an essentially east/west political axis to a north/south axis where concerns centre on the protection of economic interests, the possibilities of proliferation of weapons of mass destruction, and the growing stake which European Mediterranean states perceive themselves to have in the domestic stability of North African states. The continent may also be witnessing the radicalisation of European Islamic peoples, driven partly by external powers (Iran at the moment), but also arising from the course of the war in Bosnia, where the plausible—though as yet unproved—perception is growing that Europe is becoming significantly more anti-Islam within its own boundaries.

Such 'geopolitical scenarios' are compounded by a number of what may be termed 'overlaying factors' which are not, in themselves, security threats, but which can exacerbate other threats, act as a trigger to them, or increase their severity. Many decision-making systems can cope with a single crisis reasonably adequately, but few can cope with multiple crises simultaneously. The problem raised by such overlaying factors is that they make multiple, simultaneous crises more likely. Such factors include the chronic poverty of parts of central and eastern Europe which creates political instability in itself and reduces the stake which governments may feel they have in being part of united international diplomatic initiatives. Even states which are not suffering chronic poverty, such as Poland, the Czech Republic and Hungary, whose immediate economic prospects are quite favourable, know that their economic development is fragile and could easily be knocked off course by crises which prevented the renewal and rebuilding of economic infrastructures. Closely allied to poverty and economic disruption is the growth of international crime, narcotics and terrorism throughout eastern, central and western Europe, which—in the case of eastern Europe in particular—has the potential to undermine governmental authority to a catastrophic extent, and to increase ethnic tensions (as it has undoubtedly done between Russians and Chechens, between French and Algerians), encouraging states to close their borders and/or resort to draconian internal security measures. To this must be added the possibilities of environmental stress which may take the form of trans-border pollution, disputes over scarce resources and the increasing likelihood of major nuclear accidents among the 25 suspect reactors operating in eastern Europe (15) and Russia (10). Since all these reactors are regarded as structurally unsafe, a nuclear accident on the scale of Chernobyl is a distinct possibility in the foreseeable future.

None of these overlaying factors in themselves need be politically significant but, taken together, they can attack the essentially stable foundations of European security in the event of short-term and immediate crises. In a Europe of 45 states (defined as the area from Ireland in the west to Kazakhstan in the east and excluding North Africa and Turkey) only 18 states have a largest ethnic group consisting of over 90 per cent of the total population; in five states the largest ethnic group is below 55 per cent of the total population and in 22 others it is in between those high and low figures.[11] In mature and settled societies such calculations should not be important; but in times of short-term crises and instability they tend to loom all too large.

The implications for Britain

None of the scenarios outlined above constitutes a threat to the British homeland. At most, each constitutes a direct—and often an indirect—challenge merely to legitimate British interests. Unlike Germany, or France, for example, Britain does have a feasible option of complete non-involvement in crises such as these if they do occur.[12] It has fewer economic interests in central and eastern Europe than most of its European Union partners: British trade with Central Europe remains less than 5 per cent of its total, and investment levels are well below those of the other major west European states. It will come under far less immigration pressure which tends to be one of the ripple effects of European crises, and its own ethnic make-up—reflecting as it does new Commonwealth, rather than European, immigration—is unlikely to act as a catalyst of instability as could be the case in other European countries.[13] However conflictual the European continent is, if it is fragmented in its conflict then Britain has the option of reverting to its traditional historical role of standing aside and only becoming truly involved if a single power threatens to become dominant. This would constitute a traditionalist approach to British security policy for the future. The argument would be that as a post-colonial power in a post-Cold War world, Britain no longer has global responsibilities of any significant sort, that there is little it can do in security terms to affect the crises of European security taking place in central, eastern and southern Europe, and that non-involvement is an attractive economic option, allowing Britain to capitalise upon its virtues as a trading nation, a major foreign investor, and a dominant base for inward international investment into Europe.[14] The traditionalist argument accepts that there would be a foreign policy price to pay for effective non-involvement, such as the damage to NATO and to Britain's standing amongst its European allies, its partners in the EU, and with the United States. But such a price, the argument runs, is surprisingly moderate, since it is paid in the declining currency of the Cold War and the benefits to Britain of cuts in defence expenditure, and the greater flexibility of economic action that would follow the release from commitments would more than outweigh any loss in prestige or diplomatic weight. The annoyance of our European partners, who do not have the option of non-involvement, would not confound the economic logic of trading and investing in Britain if that logic is sufficiently strong.

This chapter argues, however, that non-involvement is a false hope, and simplifies the potential costs to an unjustifiable extent. This argument rests on the premise that, though there is a large measure of discretion in the degree of involvement Britain may feel it prudent to adopt, the option of genuine non-involvement is simply too dangerous and expensive realistically to countenance. This point of view rests on a series of arguments which run from the general to the particular, to the effect that although security problems may be physically some distance from the British homeland, they should—paradoxically—matter more to Britain now than would ever have been the case during the Cold War.

The most general argument is that the principle of common security is a long-term interest for a state as dependent as Britain on world order and stable investment and trading relationships. A failure to address security problems, and to attempt to uphold some principles of world order, particularly in Europe, in the long term weakens those principles in general. Already the Bosnian war has established a most dangerous precedent for European security whereby Britain and other powers have accepted in Bosnia that European boundaries can be changed permanently by force and have tolerated the practice of ethnic cleansing. In recognising the new Bosnian state in 1992 and then working to bring about a compromise between the Bosnian Serbs who would partition the state and the Bosnian government who have tried to uphold it, the European powers have only half-knowingly connived in breaching two of the most important principles of European security that had been successfully upheld for 48 years. This is in direct contravention to the United Nations charter and the Final Act of the Conference on Security and Cooperation in Europe (CSCE) of 1975; both of which have been diminished as international security institutions by the Bosnia war.

Apart from the question of principle, crises in Europe have ripple effects in creating flows of refugees, economic disruption and in having disproportionate effects on those states enforcing international sanctions. Most of these ripple effects are more severe on Britain's allies than on Britain itself. But this is not an argument for non-involvement since those states who are disproportionately affected by the fall-out from European crises may conclude that a failure to address problems collectively leaves them no choice but to break ranks and act unilaterally. If the European Union and/or NATO and the Western European Union are unable to take effective collective action in the face of security challenges which directly affect some of the member states, then the temptation towards unilateral action to alleviate the most immediate ripple effects of crises may be overwhelming for some major European powers. This would undermine the collective European approach to security which has been a major British interest since 1947 and would weaken NATO, which has proved an excellent vehicle for British diplomacy since its inception. More especially, a breakdown in collective approaches to European security may not only deprive Britain of a longstanding diplomatic advantage, but could undermine much more severely the essential stability of the western security community, which has existed since the mid-1940s.

Germany, in particular, is affected more quickly and more deeply than most of its European partners by crises in eastern and central Europe. France, and to a lesser extent Italy, are similarly affected by crises in the Mediterranean basin. The danger for Britain is that a Germany or France exasperated with its partners for their lack of resolve, having to cope individually with crises on or near their own borders would not only act unilaterally but would enter into bilateral arrangements with states who are party to the given crises. Such behaviour would fundamentally undermine western international institutions and would run the severe risk of 'clientism' where outside powers dabbled in crises on their doorsteps by backing different parties. A decline in collective approaches to European security, in other words, would not only weaken British diplomacy, and undermine a longstanding British foreign policy objective, but runs the risk of being severely destabilising in itself.

Finally, a similar argument applies to Britain's relations with the United States and within the United Nations. Many who are sceptical of our involvement in European crises look to the British relationship with the US and prominent role within the UN as part of the alternative formulation. But if Britain retains a strong foreign policy interest in keeping its permanent membership of the UN Security Council, and an equally strong foreign policy interest in maintaining some sense of partnership with the United States, then the implication is that it should be capable of playing a prominent role as one of the major four or five states in a Europe of 40-odd states—precisely as a contribution to its important relationship with the UN and/or the United States. If Britain lacks the means or the will to contribute to the security of its own continent and contribute collectively to that of its close allies and economic partners, then its commitment to any global conception of security must be doubted.

Such arguments have been recognised explicitly in the 1995 Defence White Paper which represents something of a departure, at least in emphasis, from those of the last three years. It begins by stating, as logically it must, that British defence policy is designed to maintain the freedom and territorial integrity of the United Kingdom and its Dependent Territories (of which there are still fourteen). But it also makes it clear that it is designed to create and preserve 'the conditions of peace and stability, within which we can pursue our national interests'.[15] There is nothing novel in this commitment: it is the 'Defence Role Three' first articulated in the 1992 Defence White Paper.[16] But it is given a prominent position in the 1995 formulation which goes on: 'Our defence policy is designed to support this wider security policy...the armed forces are being put to a broader range of uses in promoting our security interests...defence policy is interleaved to a greater degree than in the past with foreign and economic policies in the pursuit of our security goals; and the use of the armed forces is therefore more likely to be orchestrated with other instruments'.[17]

Conclusion

Britain is in a safe territorial position which is historically unique in its modern history. The post-Cold War world seems to have offered us a golden opportunity to relax. We are not ourselves threatened, and the long-term prospects for peace and stability in Europe look generally favourable. Britain has many economic interests, of course, some of which could be regarded as vulnerable in a number of ways. But in the contemporary world this is not primarily a politico-military question. 'Economic security' can only be achieved by economic means. The relevance of military force to protect economic interests—even relatively tangible economic interests such as fisheries or trade routes—is fairly peripheral in the modern world. At least in the traditional sense of the term, Britain can view its security interests with a degree of relaxation and intellectual detachment.

This must, however, be interpreted carefully. History has not bequeathed Britain a fortress within which we can sit and only emerge when we choose. Rather it has given us something that might be better characterised as a democratic and economic 'force-field' which is powerful yet also intangible and which could simply wear off if it is not properly maintained. In particular, the short-term challenges which European security is likely to face could be mishandled both by Britain or the other major powers in European politics; a series of unilateral approaches to security followed by a renationalisation of defence policies throughout Europe is not impossible. Long-term optimism, in other words, could still be undermined by short-term mismanagement which diminishes collective or common approaches to security and spirals out of control to destroy some of the essential principles upon which Europe's peace and ultimately its prosperity are based. However 'security' and 'Europe' are defined in the coming years, Britain's interests cannot be sensibly divorced from them.

To accept the argument that non-involvement is not a feasible option for Britain, however, raises two important paradoxes for British security policy. First, to uphold a collective approach to European security implies in the present environment that Britain will accept a greater involvement in local European crises and maybe conflicts which are a long way away. British security interests in the challenges that Europe now faces could therefore be described as 'deep' but nevertheless 'indirect'. This in itself is a difficult circle for policy-makers to square. Second, this may be even more difficult to square in the domestic sphere. Just war doctrine postulates that military force be used only as a last resort, but in the new circumstances of post-Cold War Europe, military action may be required early in a crisis. If military force is to be used in a more constabulary fashion then—as in the case of domestic policing—it is most effective when it is able to be used lightly, early and often. In the international sphere, however, this can look to public opinion to be precipitate and interventionist. On the other hand, where policing action by military forces *is* regarded as a last resort, then it is used in situations that have already become dangerous and is likely to require heavier forces in order to apply significant coercive power. To a domestic audience this may look risky and expensive; a slippery slope to a military quagmire.

Then too, the public understands that collective or common approaches to security should not only (or even primarily) be concerned with the deployment of military force. The most effective approaches to security are those which do not have the need to employ force. Nevertheless, even if Europe as a whole is generally successful at approaching security in non-military ways, the fact remains that Britain is one of the very few countries in Europe which possesses the capacity to deploy military forces quickly and efficiently, and has rather less direct economic muscle than its major partners to apply in preventative action to deal with security issues before they turn into crises. It is entirely possible that the strongest suit in Britain's diplomatic hand will remain its military prowess and that this will be seen by its allies as its most useful potential contribution to European security. But to a generation brought up to understand that the deployment of British forces on the European mainland was only ever part of a strategy of national survival in times of emergency, a reassessment of this perception may not come easily.

PART 2

DOMESTIC PRIORITIES

2 ROLES, MISSIONS AND RESOURCES

David Greenwood

Conservative complacency

Elected in 1979, the Conservatives remained in office through the final 10 years of east—west confrontation and well into the second half of the post-Cold War 1990s. In the latter period they first performed an initial examination of 'Options for Change' in the national defence effort to bring it into line with transformed strategic circumstances (1990–91). Then they refined their prescriptions (1991–92). Next, they developed a fresh formulation of the purposes of defence policy, together with a revised categorisation of the roles of the armed forces, an enumeration of MilitaryTtasks (or missions) in different 'policy areas' and an attribution of combatant 'force elements' to each (1992–93). All of these were elucidated in the *Statement on the Defence Estimates 1993*, which bore the subtitle *Defending Our Future*.[1] Finally, having got the front line to its liking, the Ministry of Defence (MoD) conducted an extensive Defence Costs Study aimed at achieving economies in the support area (1993–94). This latest and most radical attempt to find ways of 'trimming the tail without blunting the teeth' of the Services was described in a policy document entitled *Front Line First* issued in July 1994. Its recommendations are now being implemented.[2]

Given the foregoing, it is a fair assumption that the Tories had the defence programme broadly the way they wanted it in 1995; and, with financial projections for the next four years slowing the downward trend in real military spending—more or less continuous since the mid-1980s—they presumably had a defence budget that they were comfortable with as well.[3] So the Conservative Party entered the electoral lists committed to letting the programme-in-being run its course, while promising to manage the budget wisely.

Not that Mr Major's ministers had neglected to think about the further evolution of policy. Quite the contrary: in his last year as

Secretary of State for Defence, Malcolm Rifkind made a series of high-profile speeches outlining a prospectus for both transatlantic and pan-European security arrangements in the years ahead.[4] On the first Mr Rifkind aired a bright idea of his own—for a new Atlantic Community—which he elaborated in NATO's house journal.[5] On the second he was dull and rather predictable—extremely cautious about welcoming central Europe to the west European security community and most insistent that, within the European Union, the emphasis must be on popularly-supported, *inter-governmental* effort aimed at 'ensuring co-operation and harmonisation *where essential*'—and entirely consistent, in this last respect, with the tone of the Government's Memorandum on its Approach to the Treatment of Defence Issues at the 1996 Inter-Governmental Conference.[6]

Interesting though Mr Rifkind's reflections were, however, they contained no indications as to what the MoD's forward thinking about security *policy* implied for the country's defence posture and military provision. The obvious implication was that in ministers' eyes—and, let us assume, those of all but a handful of their backbenchers—present plans were fine. They were the outcome of a five-year-long endeavour to devise an appropriate and sustainable defence effort for the new strategic environment, and what the armed forces (and the armaments industries) now needed was a period of stability. So the MoD's current programme should be allowed to run its course to the end of this century and into the next. It might be necessary to adjust force levels or alter the force structure as time goes by, but for the business of 'defending our future' the size and shape of the armed forces, their equipment and deployment are essentially as they should be.

This is not an indefensible position, provided one accepts the premise that the programme-in-being is 'appropriate and sustainable' as claimed. However, some serious questions arise on the first count, especially when the 'five-year-long endeavour' is examined closely. Anticipating a theme to be developed later, the Conservatives did not first clarify policy objectives for the post-Cold War world, then define Service roles and missions, and finally consider required capabilities. Rather, they made early decisions about force reductions (1990–91) and, later, provided a validating policy formulation and role categorisation (1993). Furthermore, the basis of the subsequent enumeration of military missions and the attribution of front-line forces to them was not 'what *should* we now be doing'. It was 'what *are* we doing' and how are the various things that the Services do related to the new goals and what specific assets—warships, field force formations, air squadrons—serve each. Needless to say, the practical constraints on policy-making and planning rarely allow decision-makers to follow the ideal progression implied here. You start from where you are, and you must make the best of what you've got. What is not admissible, though, is to assert that, either way, you get an equally well-tailored defence effort. The MoD's current programme reflects a thoughtful reshaping of the nation's military dispositions. But it may not be entirely appropriate for present and likely future circumstances, and it would be astounding if it were the most appropriate pattern of provision for the years ahead.

As to whether the forward programme is sustainable, scepticism is in order on this count too. Anticipating a second theme to be taken up later, it is not at all certain that the resources allocated to defence in present public spending plans will actually be forthcoming; and, even if they are, it is not at all clear that they will be sufficient to pay for everything in the programme. Military expenditure after inflation may not be allowed to 'bottom-out' towards the end of the 1990s as currently envisaged. The price-level assumptions underlying the MoD's provisional cash allotments are extremely optimistic anyhow, implying a possible 'volume squeeze' on defence in the not-too-distant future. And realising the savings promised by the *Front Line First* exercise may yet be more problematic than the Ministry supposes.

Thus the air of complacency that accompanies Tory accounts of 'Options...' and after is hardly justified. Nor is the presumption that working to the Conservatives' blueprint is the best way—devotees might say the only way—to ensure that the country's defences will be in good shape, and the right shape, as we approach the millennium. This makes it all the more unfortunate that the principal opposition party appears to have no coherent alternative blueprint to offer, but has contented itself—for longer now than one cares to calculate—with a one-dimensional critique. Labour spokespersons assert, repeatedly, that a comprehensive 'defence review' is needed; but they decline to say exactly what such an exercise might involve or what it might accomplish.

And Labour's confusion

The Labour Party has painted itself into a corner in this regard. Since 1979 successive Conservative governments have performed a number of defence reviews. The major scrutiny of the defence programme that Secretary of State for Defence John Nott conducted in January–June 1981 was the first.[7] The examination of 'Options for Change' was the most recent, and the only strictly comparable undertaking. Between these two episodes, though, there were a number of occasions when ministers had to take a closer look at the balance between commitments and resources than this issue would normally attract in routine consideration of the MoD's programme and budget.[8] (I refer here to the annual scrutiny that takes place (a) in-house when the Ministry updates its Long Term Costing (LTC); and (b) Whitehall-wide when the Public Expenditure Survey Committee (PESC) looks at all spending departments' plans.) Moreover, since 'Options' there have been a further couple of important re-examinations of provision to bring the programme into line with the budget, as already noted. One was the 'Options 2' exercise which was done following the 1992 General Election and that year's savage PESC 'round'; the other was the Defence Costs Study, following a second successive PESC 'round' in which forward funding for defence was cut back sharply. Despite all this activity, however, when any issue arises related to defence policy-making, planning, programming or budgeting, Labour's response is invariably the same: the Tories have got it wrong (whatever 'it' is) and the only sensible course is to have a 'proper' defence review (whatever that means).

Invitations to elaborate on this universal prescription, often smartly evaded, do occasionally evoke a reaction. But it is not often a confidence-building pronouncement. Sometimes the 'line' is that there is a need to take stock, the Conservatives having been too busy making policy on the hoof to do this. Sometimes, when the government clearly has thought about what it is doing, the argument is that the deliberations—and hence the outcome(s)—have been 'Treasury-led' not 'strategy-led', and so there can be no possible merit in them. Sometimes platitudes have been offered, to the effect that we must 'sort out' our foreign policy, formulate a clear and consistent security policy and then, once all this has been done, address defence matters—a litany that serves only to demonstrate how long it is since Labour last held office.

Such elaborations simply underscored the fact that the all-purpose call for a root-and-branch defence review had become a virtual incantation, the easy response to any 'What would *you* do?' challenge and not a reflection of searching critical analysis of Conservative decision-making, still less original, distinctive and practical Labour thinking on how to put Britain's defences in better order. The Tories had 'taken stock' half-a-dozen times since 1979. They had to revise plans under financial pressure on several occasions; but it is manifestly wrong to dismiss all their efforts to reshape the military as ill-considered penny-pinching devoid of any strategic rationale. (Criticism of the original 'Options for Change' prospectus in such terms is particularly wide of the mark.) As for the platitudes, when Labour's spokespersons resorted to these they simply highlighted the party's lack of imaginative yet concrete proposals for resolving the permanent commitments-versus-resources dilemma in military affairs. (Moreover, to the extent that they create the impression that Labour's top people do not really know what they would do in this policy area, they foster a perception that could yet prove an electoral handicap. Ambiguity about what the party stands for now means that attitudes may be shaped by selective recollection of what it used to stand for.)

So Labour has a problem. It must invest the notion of a 'proper' defence review with some substance: partly so that, if elected, leaders who would be politically bound—some might say honour-bound—to initiate such an exercise might know what they would be seeking to achieve; partly so that a clear agenda can be developed, and disseminated, to maximise their chances of having the opportunity. It is instructive to consider how this might be done, bearing in mind some general truths about defence decision-making and taking into account some particular values that a Labour administration would doubtless wish to promote.

Under the 'general truths' heading the most important are 'you start from where you are' and—at least to begin with—'you have to make the best of what you've got.' To put this more formally:

Any incoming government inherits its predecessor's policy formulations, mission statements and mission priorities; a defence programme-in-being and associated budgetary projections; an existing force structure and force levels; equipment (in service, in production and under development) and personnel (on varying terms and conditions of service). It also takes over much else, from external treaty obligations to domestic contractual commitments.

Any incoming government, looking over its legacy, is well advised to do so systematically, considering first which of its predecessor's 'policy formulations, mission statements and mission priorities' appear robust (and which do not), then at the 'programme-in-being and associated budgetary projections' supposedly founded therein, then at the force planning questions. It should be particularly attentive where inconsistencies or incompatibilities among policy, posture and provision are apparent.

Limited freedom of manoeuvre does not mean powerlessness, however. The incoming Labour governments of 1964 and 1974 both found it impossible to reshape the defence effort to the extent and at the pace that they would have liked. Defence Ministers Denis Healey and Roy Mason were able nevertheless to make their mark.

Regarding the 'particular values' that a Labour government entering office in the late-1990s might be disposed to promote, two observations are in order.

- A Labour Cabinet would doubtless wish to find additional resources for selected civil purposes—the health service, education, public infrastructure projects, tax breaks to encourage industrial investment and training, or whatever—but would probably not wish to cut back sharply on the rest and would certainly not wish to increase the overall tax burden (or public sector borrowing). It might, therefore, favour either early reductions in the MoD's target estimates or capping defence expenditure, perhaps at the election-year level (much as the Wilson government of 1964 did).[9]

- At the same time, there is no doubt that Labour would wish to preserve a defence programme *appropriate* to post-Cold War conditions and *sustainable* in the light of the general state of the public finances on the one hand, its own most urgent priorities on the other.

The reappearance of those words 'appropriate' and 'sustainable' in the second observation is a clue to what would-be Labour ministers ought now to be considering as a possible agenda for their 'proper' defence review.

The explicit policy commitment in this area should be that, given office, they would amend neither fundamental national security objectives nor the MoD's characterisation of the armed forces' essential post-Cold War roles. But they should take a long and hard look at the extensive catalogue of military tasks upon which the programme-in-being is based, at mission priorities, and at alternative ways of discharging some responsibilities. They would do so for the simple reason that the appropriateness of certain tasks, priorities and forms of provision is not self-evident. No less important, they should take a long and hard look at the budgetary projections associated with the programme, for the simple reason that the sustainability of the current defence effort is questionable too.

Roles and missions

As hinted earlier, there is a sound rationale for such scrutiny because of the way in which business has been done at the MoD since 1989/90. The 'Options...' inquiry begun then was driven by two impulses: a felt need to respond immediately to changed strategic conditions on the continent and an urgent requirement to reduce the MoD's forward spending plans. There has been much sterile argument over which was dominant: was the exercise 'strategy-led' or 'Treasury-led'? The answer is neither, or both, according to taste. The end of the Cold War allowed programme changes to be made that conveniently relieved the pressure on the MoD's budget. Faced with the problem of making ends meet, the Ministry decided on big force reductions, the burden of adjustment falling on British forces in Germany (because of the extent to which the threat there had receded). The main decisions about reductions had been made by the middle of 1990—and were announced then—even though the details were not spelt out until the following year (in the 1991 Defence White Paper, subtitled *Britain's Defence for the 90s*, a hostage to fortune if ever there was one).[10]

A period of relative tranquillity ensued in the run-up to the 1992 General Election. It was abruptly ended, however, by that year's (post-election) PESC 'round' in which the resource allocation to defence for the current financial year (1992/93) was cut sharply, as were those for following years. In consequence, major changes to the 'post-Options' programme had to be made. At the MoD they went back to the drawing-board and conducted 'Options 2'. The result, predictably, was further force reductions.[11]

Somewhere along the line here—probably after fighting an early losing battle with the Treasury—a smart official (or officer) at the MoD must have realised that the time had come for a restatement of the purposes of the defence effort. For a decade—since Nott's *Way Forward* in fact—these had been expressed in terms of the 'five pillars' of policy: maintenance of a strategic nuclear force, protection of the home base, a contribution to NATO's order of battle for land-air warfare in Europe, a contribution to Alliance forces for maritime operations in the eastern Atlantic and Channel, and the fulfilment of residual extra-European obligations. These were essentially threat-based formulations, devised for a context of east-west confrontation. They provided a poor basis for demanding taxpayers' money when 'the threat' had gone, the Cold War was over, and expectations of a 'peace dividend' were widespread. So the Ministry offered, initially in the 1992 Defence White Paper, a new statement of objectives—'to contribute to maintaining the freedom and territorial integrity of the United Kingdom and its dependent territories and [the country's] ability to pursue its legitimate interests at home and abroad'—and a new capability-based framework for the armed forces' responsibilities, with the enunciation of three (overlapping) Defence Roles, *viz.*

1. '...to ensure the protection and security of the United Kingdom and our Dependent Territories *even when there is no major threat*' (emphasis added);
2. '...to insure against a major external threat to the United Kingdom and our allies' (through NATO membership and the provision of forces to the Organisation's military structure);
3. '...to contribute to promoting the United Kingdom's wider security interests through the maintenance of international peace and stability'.

That was what the MoD would want funds for in the new environment, and it would need all of the £20-odd billion per annum in the expenditure plans.[12]

Over the next 12 months the ministry carried this process a stage further. It developed what the Secretary of State described as a 'new and fundamental analysis of the way in which defence assets *are*, and will *need to be*, employed' (emphasis added): in order, according to the Minister again, to expose 'the links between policy, the tasks which our forces are called upon to undertake and overall force structures'; and in the hope, he might have added, thereby to strengthen the defence hand in future budgetary battles.[13] Put briefly the MoD enumerated the various Military Tasks (missions) that the armed forces were doing, or were expected to have the capacity to do, at this time; and each mission or responsibility was assigned to one of the three Defence Roles. Further, the planners looked at all the force elements on the Services' books—warships, field force formations, air squadrons—and allotted each asset or unit in this instance not to a single Role but to all relevant Roles (multiple earmarking).[14] The enumeration of Military Tasks yielded a 50-item checklist (reproduced as an Appendix to this Chapter) and the 1993 Defence White Paper—*Defending Our Future*—described in detail the provision currently made for each. The same document showed the (typically multiple) attribution of force elements to Roles plus a rough-and-ready calculation of the incremental cost of each.[15]

The material describing all this in *Defending Our Future*, and especially the tabulated results of the work, summarise what was obviously a formidable intellectual and clerical effort. The data also represent a valuable contribution to transparency in defence policy-making and planning. The MoD has forged a useful tool for itself and provided much useful information for the political and analytical communities.

For present purposes, however, other features of the operation need to be highlighted. In the first place, the checklist of Military Tasks (see Appendix p.281–90) sets out what the armed forces *are* doing and how their activities may be said to serve the newly-stated policy purposes. It is not a presentation of the jobs the Services *should* be doing, drawn up following an exercise in mission specification (and mission priority-setting) taking the Defence Role formulations—which are unexceptionable—as its starting-point. In the second place, the brief accounts of current provision for each function (*Defending Our Future*, Chapters 3–5) are just that: statements of what we *are* fielding, with no reference to alternative scales or patterns of provision that *might be* equally capable of fulfilling the requirement. In the third place, the assignment of force elements to Roles is a formal—and, on the MoD's own admission, somewhat arbitrary—allotment of the national order of battle as it now *is*, and in no way precludes the possibility that a quite different force structure *could be* no less effective in fulfilling those Roles. In short, path-breaking though the MoD's 'new and fundamental analysis' unquestionably was, it was an exercise in enumeration, description and attribution—neither more nor less.

The other important comment on this account of 'how business has been done at the MoD since 1989/90' is, of course, that both enunciation of the new policy framework, and the pragmatic 'enumeration, description and attribution' to complement it, came *after* the key decisions about force reductions had been made. Choices about the size and shape of the armed forces were made first, the validating rationale followed. In view of this, and the questions just raised about the MoD's elaborate demonstration that what had been done nevertheless made sense, no new government entering office in the mid-1990s—of whatever stripe (including yet another Conservative administration)—should regard the defence programme it inherits as uniquely appropriate for the post-Cold War world.

In particular the case for a long and hard look at the Military Tasks (MTs)—enumerated in the Appendix (p.281–90)—is obvious, and not only in the light of the checklist's provenance.

- Whether the United Kingdom *should* be devoting resources to maintaining an independent strategic and sub-strategic nuclear capability in present circumstances ought at least to be put to the question, especially since in the lifetime of the next government the question of Trident successor systems may arise (MT1.1 affecting MTs 1.2 and 1.3). (This is also discussed in Chapters 16 and 18.)

- Whether specific provision for Military Aid to the Civil Power in Northern Ireland should continue to be made ought similarly to be questioned. (Should not MT.1.4 'absorb' MT 1.5 soon?) The scale of provision is bound to be an issue anyhow, if the 'peace process' progresses (with implications for the size of the infantry).

- The scale and pattern of provision under several other Defence Role 1 headings merit examination, from the arrangements made for policing the waters and airspace around the British Isles to the protection of remaining distant dependencies (MTs 1.9 *et seq*.).

- What the United Kingdom *should* contribute to NATO in the later 1990s and beyond ought to be addressed as well. It is generally agreed that the transformation of the Organisation and the recasting of its military dispositions still has some way to go. Indeed a further strategy and force structure review may be launched before too long, the first post-Cold War exercise of 1990–91 having been clearly overtaken by events. Certainly the designation of Immediate Reaction, Rapid Reaction and Main

Defence Forces (plus augmentation forces) ought to be abandoned shortly in favour of a looser structure of military capabilities organised with the composition of 'adaptive force packages' in mind, for deployment in contingency operations as Combined Joint Task Forces (CJTFs). *Who* should contribute *what* to the 'new model' NATO—possibly an enlarged NATO—will then have to be considered, with issues like transatlantic and intra-European burden-sharing, options for a 'division of labour' (and task specialisation) and the practicalities of ensuring that there are 'separable but not separate' forces available for operations under the aegis of the Western European Union—all entering the reckoning.[16] In short, the size and nature of the United Kingdom's contribution to collective or joint security 'insurance' is likely to require consideration again in the not-too-distant future (MTs 2.1–2.13).

- A blurring of the distinction between Defence Role 2 and Defence Role 3 (in the MoD's formulation) will occur if NATO does indeed transform itself completely from an *alliance-in-being* to deal with 'the threat from the East' to a looser *coalition-in-waiting* for diverse contingencies in and around Europe (or further afield), especially if the peacetime organization, stationing and training of forces are adapted with realization of the CJTF concept in mind. While the distinction is blurred already so far as the provision of 'force elements' is concerned—by the multiple earmarking technique (which as the MoD practises it means that no capabilities are expressly maintained for Role 3)—the question arises: if a 'new model' NATO looks to the United Kingdom for *fewer* forces than today, would the nation nevertheless wish to maintain an independent national intervention capability with all-round competence? (MT 3.1—3.5, also affecting 3.6 and 3.7.)

There *is* an agenda for a defence review here. As noted earlier, the Conservatives' restatement of defence policy objectives ('to contribute to maintaining the freedom....etc.') and even their three-fold role categorisation are unexceptionable. But the appropriateness of certain tasks, mission priorities and forms of provision is not self-evident. Tory complacency about the programme-in-being is misplaced. Labour's confusion about what it would actually do if given office need not persist: there are numerous questions to be asked about current provision and present plans.

Resources

The sustainability of the MoD's programme is in doubt as well. On the face of it, this ought not to be the case. In the 1994 PESC 'round' the Ministry escaped further assaults on its funding. Its cash appropriations fell, but that was for technical reasons—to be explained presently—including revised assumptions about likely inflation. On those assumptions, as it happens, real defence spending should level out soon. However, it is *not* certain that the Ministry will get the cash in its provisional appropriations: there is, after all, a General Election looming. Nor is it certain that prices will be held at the ultra-low level allowed for in them.

The official financial prognosis is clear. For 1995/96 the MoD has a cash appropriation of £21.7 billion. The target estimate for 1996/97 is £21.9 billion. That for 1997/98 is £22.3 billion. Provided inflation is held in check, these budgetary allocations imply only a slight diminution in real spending: from £20.6 billion (at 1993/94 prices) in 1995/96 to £20.3 billion in 1996/97, slipping to £20.2 billion the year after. This opens up the prospect of expenditure after inflation 'bottoming-out' around the end of the decade, with annual outlays equivalent to £20 billion at these same 1993/94 prices. This is the essential message of the latest figures presented in the *Red Book* published in November 1994 (hereafter *Red Book 94*). It is a message that has not been widely appreciated, largely because the defence numbers are markedly lower than in earlier projections. The difference, though, is attributable to, first, an important coverage change affecting these numbers specifically; and, secondly, the Treasury's revised estimate of inflation in 1994/95 and revised assumptions about inflation in later years, affecting all categories of public spending.[17]

To elaborate on this a little: the defence expenditure data for budgeting in 1994—that is, the outcome of PESC 93, recorded in the *Red Book 93*—was as follows:

Table 2.1 Defence expenditure data for budgeting 1994, from PESC 1993

	94/95	95/96	96/97	97/98
£ billion (cash)	23.5	22.7	22.8	(22.5)

(The final figure is in parenthesis because it is a guesstimate.)

This cash provision incorporated some 'extraordinary' items, for the continuing costs of recovery from the Gulf conflict and for redundancy plus other expenses associated with transition to the (amended) 'post-Options' defence programme. But these amounted to only £0.1 billion in 1995/96 and half that in 1996/97. So far as budgeting for 1995/96 and after is concerned, therefore, this was funding for a 'baseline' post-Cold War and post-Gulf War defence effort, free of 'extraordinary' allotments.

Expenditure plans were, however, revised from 1 April 1994 to take into account (a) the transfer to a new Cabinet Office Vote of Provision for the Security and Intelligence Services and (b) a few minor transfers of provision between the MoD and other departments. These coverage changes reduced the resource allocations for the MoD's programme by £0.6 billion or thereabouts each year. The budgetary projections for 'present plans' that formed the point of departure for the 1994 PESC 'round' were, therefore, as set out here.

Table 2.2 Budgetary projections for 1994 PESC round

	94/95	**95/96**	**96/97**	**97/98**
£ billion (cash)	22.9	22.1	22.2	(21.9)

(The final figure is in parenthesis, as before, because it is not an official target estimate.)

The first of these sums was, of course, the amount already made available to the MoD as its *budget* for the fiscal year (for spending as detailed in the 1994 Defence Estimates). The numbers for the following two years were the new *target estimates* for those years, reflecting changed coverage.

In the United Kingdom public expenditure is planned on a cash basis. Deciding the financial provision for a spending department in a given year thus involves gauging the 'volume' of resources that will be required to sustain its programme of activity, costing that requirement at the prevailing price level *and* incorporating an allowance for any rise in prices expected in, or by, the given year. If insufficient allowance is made under the third heading—if the prices that the particular department has to pay for its inputs turn out to be higher than anticipated, but its cash limit is nevertheless enforced—then it experiences a 'volume squeeze' on its programme. Cutbacks are necessary to stay within budget. Conversely, if too generous an allowance is made—if the department finds it can buy its inputs for less than expected, but still gets its appropriation—then it enjoys a windfall 'inflation bonus'. It can do more with its budget than it had thought it would be able to do.

Over the years the MoD, like most other spending departments, has experienced the 'volume squeeze' more often than it has enjoyed an 'inflation bonus'. At the start of the 1994 PESC 'round', however, the Department had a £22.9 billion budget, predicated on a general rate of inflation in 1994/95 of 4 per cent, but prices in the economy were not rising at anything like that rate. It was looking at a (revised) target estimate for 1995/96 of £22.1 billion, predicated on (further) inflation in that year of 3.75 per cent, while all the indications were that prices would not go up to anything like that extent. The prospect was, of course, too good to be true. Having realised that it had got its inflation expectations seriously wrong, the Treasury promptly decreed that it would be lowering the price-rise allowance for 1994/95 from 4 per cent to 2 per cent and that for 1995/96 from 3.75 per cent to 3.25 per cent, in line with hastily revised forecasts. (It left the allowance for 1996/97 unchanged at 2.5 per cent, and told departments to assume no more than 2.25 per cent inflation for 1997/98.)

A £22.9 billion budget incorporating an inflation allowance of 4 per cent becomes a £22.5 billion budget if the allowance falls to 2 per cent. A following year's target estimate of £22.1 billion incorporating price-rise provisions of 4 per cent then 3.75 per cent becomes one of £21.6 billion if the inflation factors are 2 per cent and 3.25 per cent. Give or take a little, these are the levels to which the MoD's current-year and forthcoming-year allocations were marked down in the PESC 94 exercise, as recorded in the *Red Book 94*. The full projections are as follows:

Table 2.3 MoD's current-year and forthcoming-year allocations

	94/95	**95/96**	**96/97**	**97/98**
£ billion (cash)	22.5	21.7	21.9	22.3

(No parenthesis for the final figure this time. The fiscal year 1997/98 has now appeared on the official public expenditure planning horizon.)

The 1994/95 amount is now an 'estimated outturn' figure: there is to be no sanctioned 'inflation bonus' for the MoD. The reduced target estimate figures for the forward years are as cited earlier. There is no windfall here either, but nor does the Treasury appear to have insisted upon yet more real cuts in military provision.

Table 2.4 Defence budgeting 1993–94 (budgets/target estimates in £ billion)

		1994/95	1995/96	1996/97	1997/98	Notes
A	PESC 93 (cash))	23.5	22.7	22.8	(22.5)	Red book 93
B	SDE 94 (cash)	22.9	22.1	22.2	(21.9)	Coverage change
C	PESC 94 (cash)	22.5	21.7	21.9	22.3	Red Book 94
D	Difference B-C	-0.4	-0.4	-0.3	(+0.4)	B minus C
	of which					
	'declared change'	-0.3	-0.3	-0.2	–	Red Book 94
	inflation factor	-0.1	-0.1	-0.1	–	inferred
E	PESC 94 (at constant prices)	22.1	20.6	20.3	20.2	Red Book 94 (1993/94 prices)

The Treasury does not *appear* to have imposed further programme reductions. But this cannot be asserted with complete confidence. That is because there is one minor unresolved mystery in the official figures (arising because the declared 'change from previous plans' in the *Red Book 94* does not quite square with the adjustments just described). It looks as though the Treasury may have made provision for 'clawing back' any 'inflation bonus' that might accrue to the MoD in 1994/95, despite their best efforts to prevent it. Whatever the reason, the 'mystery' is almost certainly associated with the 'inflation factor' somehow; and that inference is acknowledged in Table 2.4 which summarises the evolution of the allocation of resources to defence from the PESC 93 benchmark to date, as traced in the preceding paragraphs.[18]

The 'bottom line' of this table is the defence expenditure at constant prices calculation given in the Red Book 94 (using 1993/94 values). This confirms that the 'volume' decline, or shrinkage, in the national defence effort appears to be coming to an end. Slowly rising cash allocations to defence are the order of the day again (Line C), as they were in the initial 'post-Options' prospectus. In circumstances of very low inflation—an assumed 2.5 per cent for 1996/97 and 2.25 per cent for 1997/98, remember—this will lead to a 'bottoming-out' in constant-price spending as we near the end of the decade at a level of real resource allocation some 10 per cent below the current one (Line E).

Is that a reasonable prospect? A 'true blue' optimist would argue that it is, noting that in 1994 the 'underlying' general inflation rate—as measured by the Retail Price Index excluding mortgage interest payments—was successfully driven to a 27-year low of 2 per cent *and* that the government appears fully committed to holding it down. So the days of the perennial 'volume squeeze' on the defence programme—and either chronically underfunded military dispositions or regular, arbitrary prunings of the programme—may be gone, perhaps for good. Further, the MoD and the armed forces have been cut down to size and are now concentrating on their 'core business' to an extent unheard of hitherto. So there will be little, if any, scope for fraud, waste and mismanagement in the future.

Having said that, a critic might fairly question just how sustainable the programme-in-being really is. Recent evidence from the National Audit Office and elsewhere suggests that while the MoD has assuredly learnt all the fashionable management-speak of the day it still frequently takes diabolical liberties with taxpayers' money. More important, holding the annual rate of 'underlying' inflation in the 2–2.5 per cent bracket is going to prove very difficult. Yet pressure for tax cuts and/or boosting civil spending will almost certainly preclude the allocation of additional resources to defence in the immediate future. So it could be back to the familiar 'volume squeeze' even if the MoD does continue to get a mite more cash year-by-year.

It has to be said that history, especially recent history, is on the side of the sceptic in this caricature confrontation. Indeed, the position in the mid-1990s was powerfully reminiscent of 1990–91. The Defence Secretary's message on presentation of the *Front Line First* paper, on the eve of the 1994 parliamentary recess, that essentially 'we have axed support, but preserved the front line' might have come from the same sound-bite mill that had given his predecessor, Tom King, a crisp phrase—the armed forces of the future will be 'smaller, but better'—to go with his initial statement on 'Options for Change' exactly four years earlier. The foundation for the latter claim was, of course, reductions in Service manpower and in warships, field force formations and aircraft squadrons of, roughly, one fifth; but a much smaller reduction in projected defence budgets. Two years later, however, expenditure plans underwent the first of two 'downward revisions' which, together, were to produce a fall in the real value of provision fully commensurate with, if not going beyond, the contraction in the national force structure

Table 2.5 The defence costs study: estimated job reductions by the year 2000

	RN	Army	RAF	Civilian	Total
HQ Reductions	150	200	500	900	1,750
Recruiting/Manning	600	900	2,200	–	3,700
Medical Support	250	250	250	270	1,020
Rationalisation of Bases/Depots	800	300	2,150	3,100	6,350
Rationalisation of Science etc.	–	–	–	1,300	1,300
Improvements in IT	20	300	100	430	850
Other Civilianisation	–	–	1,500	–	1,500
Other Support Posts	80	250	800	1,100	2,230
Total	1,900	2,200	7,500	7,100	18,700
% change	-4.1	-1.9	-11.6	-6.5	–

Source: Front Line First, Report of the Defence Costs Study (MOD), London, HMSO, 1994, p.39.

envisaged in the initial 'Options...' prospectus. The obvious question is: will Foreign Secretary Rifkind perhaps have to eat the words of Defence Minister Rifkind in, say, later 1996?[19]

The chances are that these words *will* return to haunt Mr Rifkind. There could well be an early return of the 'volume squeeze' and of level cash funding for defence. On top of that, many of the savings that implementation of the Defence Costs Study (DCS) proposals are expected to yield may prove illusory, while others may damage combat effectiveness in ways that their advocates have failed to recognise (not least through effects on Service morale and *esprit de corps* that are already evident). Some savage cuts are programmed, as the data on envisaged manpower reductions in Table 2.5 make clear.[20]

Conclusion

The significance of this last observation cannot be overemphasised. The DCS was done in order to find an annual £750 million of savings on the defence budget from 1996/97 to enable the MoD to stay within the financial straitjacket resulting from the tough PESC 93 'round' without reducing force levels or recasting the force structure yet again (in what would undoubtedly have been designated 'Options 3' by analysts and the media). Failure to realise the economies that the *Front Line First* document promises, failure to deliver the personnel cuts that Table 2.5 prescribes—these will send the Ministry's planners back to the drawing board again, even if the sums in the current projections of Table 2.2 are forthcoming.

Whether they *will* be forthcoming is in doubt. Nothing in the state of the public finances, or the Conservatives' recent standing in opinion polls, suggests that the 1995 PESC 'round' will be an easy one for any spending Department, except maybe those with what are clearly vote-winning (or vote-saving) programmes to promote (or protect). But that is not all. Even if the cash keeps coming, it may not buy all that it is supposed to buy. Those austere inflation allowances look more and more like harbingers of the dreaded 'volume squeeze'.

This is the argument that the *sustainability* of the current forward programme for defence cannot be taken for granted. As for its *appropriateness*, earlier argument has shown that Conservative complacency on this score is misplaced as well.

There was nothing wrong with the Tories' restatement of defence policy objectives for the post-Cold War world, nor with their formal redefinition of the roles of the armed forces, notwithstanding the cart-before-the-horse tactic of reducing capacity first and providing a validating rationale for the cuts later. However, the enumeration of missions (Military Tasks), the description of them, their attribution and that of their costs to major roles were done in the same spirit of legitimising existing provision: categorising what the Services were already doing, or currently had the means to do. The powers-that-be did not explicitly ask what the Services should be doing, or should have the means to do in the new strategic circumstances. This is not satisfactory. There ought to be such an examination covering the checklist of tasks itself, mission priorities, and alternative approaches to provision for them; and it should be a no-holds-barred examination, not another search for good reasons to carry on doing what is being done now, with the capabilities the country happens to possess now. This is what any 'defence review' ought to be about in the mid-1990s. This is what the Labour Party should commit itself to doing; and it should say so clearly, the sooner the better.

3 RESTRUCTURING THE BRITISH ARMY

Colin McInnes

Balance, commitments and overstretch

In the mid-1990s thinking about the Army's structure is dominated by the idea of a 'balanced force' capable of reacting to any realistic contingency. The Army argues that, because the future is uncertain, it must therefore be capable of reacting to a wide variety of possible threats—from a high-intensity/high-technology war in Europe or beyond, through to low-intensity/low-technology operations such as peacekeeping and humanitarian relief. The Army therefore faces the challenge of maintaining an all-round capability at a time of stringent defence cuts.

The Army also faces the problem of matching new commitments to dwindling resources. In the immediate aftermath of the Cold War it was widely assumed that the Army could be substantially reduced with safety since its primary commitment for the past two decades, the defence of West Germany, had all but disappeared. In the event, the Army was deployed to new commitments in the Gulf, Bosnia and Rwanda. These new commitments clearly helped the Army in the battle for resources. The Army demonstrated its utility in a variety of roles in the post-Cold War world and escaped relatively unscathed in the 1992/3 defence cuts. But these new commitments also placed additional burdens on the Army and reinforced the sense of overstretch. It is widely rumoured that Washington approached the British government informally over the possibility of sending British troops to Somalia, but that the government was forced to decline, in part because the Army was already over-committed. In addition the Army acquired the command of Allied Command Europe Rapid Reaction Corps (ARRC), certainly the most important new NATO ground force and arguably the only one relevant to the post-Cold War world.[1] But with this prestigious and politically important command comes an implicit requirement to

Table 3.1 Structure of the ARRC

1.	1st UK Armoured Division	3 x armoured brigades	
2.	3rd UK Division	2 x mechanised brigades 1 x airborne brigade	plus 1 x Italian brigade
3.	One US armoured division		
4.	One German division		
5.	Multinational Division Central—MNDC	includes 24 UK Airmobile Brigade	
6.	Spanish division	rapid reaction force	possible addition from another NATO member
7.	Italian framework division	2 x Italian brigades 1 x Portuguese brigade	
8.	Multinational Division (South) (MND(S)	brigades from Italy, Greece and Turkey	
9.	A Greek division?		
10.	A Turkish framework division?		

Note: The status of some forces is uncertain. Up to four divisions may be deployed, the Commander of the ARRC choosing those which best meet the requirements of a particular crisis.

provide it with substantial forces—indeed the British have committed more forces to the ARRC than any other NATO member state (see Table 3.1) as well as the bulk of its Headquarters staff. The result has been continued pressure on already scarce resources—although the Army has used the command of the ARRC as a political weapon in Whitehall battles to fend off further cuts.

The Army's response

The Army reacted to the perceived requirement for a 'balanced' force and to the problem of overstretch in two ways. First, forces are now 'double-hatted'—that is, they have two or more roles to play. Thus 3rd Mechanised Division is committed to the ARRC, and has also been assigned the role of independent action outside NATO, while individual units may be deployed to Northern Ireland as part of the Emergency Tour Plot. This may work in peacetime; but if a series of crises erupt, double-hatted forces may be unable to meet their full obligations. It should be noted that such crises need not be simultaneous for this problem to occur given the requirement for periods of rest and training between tours.

Second, the Army retains an emphasis upon the ability to fight a high-technology/high-intensity war, arguing that it is easier for such forces to engage in low-intensity operations (peacekeeping, humanitarian relief, etc.) than for low-technology/low-intensity forces to fight a high-intensity war. Thus the Army retains a substantial armoured corps, a comparatively large number of artillery regiments, and substantial numbers of armoured infantry, forces designed primarily for a high-intensity war (see Table 3.2), while its new equipment plans continue to focus on the high technology battlefield. The 1994 Statement on the Defence Estimates, for example, identifies the following new equipment plans for the Army:

- new main battle tank (Challenger 2)
- new medium range anti-tank missile (MR TRIGAT)
- new attack helicopter
- new air defence missiles (High Velocity Missile and Rapier Field Standard C)
- improvements to Multiple Launched Rocket System (MLRS)
- the second production phase of the Battlefield Artillery Target Engagement System (BATES)
- modern bridging equipment
- new combat radio
- Demountable Rack Off-loading and Pick-up System (DROPS)[2]

All of these are relevant to the high-tech/high-intensity battlefield; few have possible applications in low-intensity operations. The Army is therefore preparing to fight the sort of war it is least likely to encounter, on the basis that this is the most difficult task to prepare for and it is easier to adapt from this high capability to lesser tasks than the reverse.

Problems

This policy raises a number of problems. The bulk of the Army's heavy armoured forces cannot be moved from their present deployment in the UK and Germany due to the lack of strategic lift and support capabilities. Although a residual commitment to defend

Table 3.2 Combat strength of the army in 1995, as planned under options for change

A. Household Cavalry and Roya Armoured Corps (11 regiments)	8 x armoured regiments 2 x armoured reconnaissance regiments 1 x training regiment (also to be used for reconnaissance purposes) (1 x mounted regiment x ceremonial duties)
B. Royal Regiment of Artillery (16 regiments)	9 x field regiments (AS 90) 3 x MLRS regiments 4 x air defence regiments (2 x Rapier, 2 x Starstreak) (1 x King's Troop x ceremonial)
C. Corps of Royal Engineers (10 regiments)	7 x engineer regiments 1 x RAF support regiment 1 x explosive ordnance disposal regiment 1 x engineer regiment resident in Northern Ireland
D. Royal Corps of Signals (11 regiments)	10 x signal regiments 1 x electronic warfare regiment 5 x independent signals squadrons
E. Infantry (38 battalions— 31 March 1998) (increased to 40 battalions in February 1993)(Warrior)	8 x armoured infantry battalions 4 x mechanised battalions (Saxon) 2 x airmobile battalions 2 x parachute battalions 21 x general purpose battalions 1 x ACE Mobile Force (Land)
F. Army Air Corps (6 regiments)	2 x antitank regiments 2 x airmobile regiments 1 x general aviation regiment 1 x general aviation regiment resident in Northern Ireland 4 x independent flights

NATO territory remains, the threat to Germany has all but disappeared and warning times are calculated in terms of years rather than days or weeks. In contrast crises such as the Iraqi invasion of Kuwait may erupt with very little warning and require a quick western response. The Army's requirement for high-intensity war therefore is *not* comparatively large numbers of heavy armoured forces based in Germany and the UK which cannot be moved quickly

Table 3.3 1st (BR) Division, Operation Granby (Gulf War): major combat units

4th Armoured Brigade	1 x Armoured regiment (Challenger) 2 x Armoured infantry battalions (Warrior) 1 x Armoured engineer regiment
7th Armoured Brigade	2 x Armoured regiments (Challenger) 1 x Armoured infantry battalion (Warrior) 1 x Armoured engineer regiment
Artillery Group	1 x Medium reconnaissance regiment (CVR(T)) 1 x Heavy regiment (MLRS) 1 x Heavy regiment (M110) 3 x Field regiments (M109) 1 x Air defence regiment (Rapier) 2 x Air defence batteries (Javelin)

and en masse. *Rather* the requirement is for a rapidly deployable force. This will inevitably operate as part of a coalition: Britain lacks not only the numbers but also certain strategic resources (particularly satellite reconnaissance) to 'do it alone' except in instances of comparatively light opposition (such as the Falklands). *It is therefore better for Britain to have a small but highly capable force for high-intensity warfare which can be moved to where it is needed rather than a larger one which cannot be moved.*

The Gulf War provides an excellent illustration of this. Despite having three armoured divisions in Germany, the Army could only deploy a single, small division in the Gulf, with just three tank regiments, three armoured infantry battalions and five artillery regiments (excluding air defence, see Table 3.3). An additional armoured regiment—16/5th Lancers—was deployed with the artillery as a medium reconnaissance regiment in lightly armed CVR(T)s. In contrast, under 'Options for Change', the government plans for eleven armoured and twelve artillery regiments. Clearly this offers a substantial surfeit of capability for Gulf War-type operations.

More damning however is the support such forces require to fight over any period of time. During the Cold War, the Army's conventional role was limited to that of a shop window force in Germany: it looked good, but was not intended to fight for any length of time before nuclear weapons would be used. Large stockpiles of spares and ammunition could be avoided because the conventional phase of a war would be relatively short. The Gulf War however demonstrated the speed with which even a comparatively small force devours ammunition and spares. Stockpiles in Germany were plundered to dangerously low levels, and some front line equipment in Germany was cannibalised to provide spares for forces in the Gulf. The best example of this was the Challenger tank, whose power pack proved so unreliable that tanks in Germany had to be cannibalised to provide spares. There were 211 Challengers deployed to the Gulf, and 279 power packs changed during the course of Operation Granby. Most of these came from tanks in Germany.

Even with sufficient stockpiles there is the problem of moving equipment into the theatre, and then from ports and airfields to the forces in the field. UK airlift capabilities are limited. Although the European Future Large Aircraft might be capable of carrying limited numbers of armoured vehicles, the choice of a new version of the Hercules to replace at least half of the RAF's current fleet means that, for the foreseeable future, the UK will have no capability for airlifting even limited numbers of armoured forces. As a result they would have to be moved by sea, a slow process complicated by potential problems in leasing ships to take equipment into a war zone. Once in theatre forces need to be moved from ports to the combat zone, and resupplied on a constant basis. During January 1991 for example, a single regiment of the Royal Corps of Transport covered over 1.5 million miles moving supplies from the port of Al Jubail to 1(BR) Armoured Division's concentration area in preparation for the ground offensive against Iraq. Without enormous Saudi host nation support (particularly in providing fuel) such a move would probably not have been possible.

Deployment of armoured forces overseas is therefore a major undertaking, and one for which the Army remains relatively poorly prepared. In particular combat operations (and intensive training immediately prior to combat) consume vast amounts of fuel, ammunition and spare parts, stockpiles of which have not traditionally been a priority for the Army.

To sum up:

- The Army's high-technology/high-intensity forces are relatively static in terms of their deployment, and only a small percentage can be deployed and supported overseas. Given the unlikelihood of a war in Germany (and warning times of several years) is such a large armoured force required? A smaller armoured force adequately supported in terms of transport and stockpiles would meet Britain's requirements better than the large, under-resourced armoured forces currently deployed in Britain and Germany. Savings moreover could be re-directed to areas currently overstretched. A possible objection to a much reduced armoured corps is the lack of cover in Germany should such forces be required elsewhere. But the Clinton administration's Bottom Up Review and (albeit to a lesser extent) the Bush administration's Base Force both placed the US Army in Europe in this very position. US troops in Europe have been justified to Congress in terms of a forward base near a variety of trouble spots in the Middle East and Africa. A second Desert Storm would see the bulk of US forces in Europe redeployed elsewhere. This has not been seen as a problem in NATO, the US or Germany.

- The assumption that it is easier to convert forces trained and equipped for high-intensity warfare to low-intensity operations may be correct, but it needs to be qualified. First, forces for low-intensity operations often need a period of training before they can be deployed. For general purpose infantry such training may be negligible since many of the skills required would be part of their basic infantry training. But if armour or artillery are to be deployed as infantry, or even if mechanised infantry is to be deployed as 'leg' infantry, additional training may be required. Armoured troops cannot simply step out of their tanks onto the streets of Belfast or Sarajevo; forces for high-intensity warfare cannot always be quickly converted into forces for low-intensity operations. Second, during the period when such forces are redeployed in low-intensity operations, their expensive equipment is stored in warehouses—an inefficient use of scarce resources (the deployment of an armoured regiment to Northern Ireland for example meant on the order of £100m-worth of equipment being placed in storage). Third, the use of armour and artillery as infantry in the Emergency Tour Plot is an explicit acknowledgement that there are insufficient general purpose infantry battalions, while the section above demonstrated that there are more armour and artillery units than can be deployed in any scenario short of general war in Europe.
- Although the government promised a better equipped Army as

part of 'Options for Change' (the phrase 'leaner but meaner' was often used), this was greeted with a scepticism which now appears fully vindicated. The Army is in danger of becoming a hollow force—implicitly acknowledged by the Defence Costs Study which attempted to find savings in the support areas to protect overstretched and under-resourced front-line forces. Expensive, high-technology equipment is clearly required if the Army is to fight a modern, high-intensity war; but it is increasingly difficult to fund the weapon systems, spare parts, huge stockpiles of ammunition and the necessary support services for the planned level of forces. A choice must be made: does the Army need a comparatively large but under-resourced capability for high-intensity warfare, or one which is smaller but well resourced?

- Although the future is, by definition, unknowable, the policy planner needs to make certain assumptions about the future nature and incidence of Army deployments. The likelihood of the Army being deployed in a high-technology/high-intensity war is slight *because there are few other powers capable of fighting such wars*, and of these only a handful threaten British interests (the Gulf, Europe, and possibly Korea). In contrast the number of low-intensity operations are already far outstripping our ability to react. National interests, humanitarian concerns and maintaining Britain's international standing all mean that the Army is likely to continue to be deployed in reaction to such crises. But the Army's emphasis upon the ability to fight a high-technology war in an era of financial stringency inevitably means that there are opportunity costs in not being also able to provide sufficient forces for low-intensity operations.

- Armoured forces require substantial areas of land on which to train, and are extremely destructive of land when training. For the past 50 years, most armoured training has been done in Germany and at the BATUS facility in Canada. With the withdrawal of forces from Germany however, and pressure in Germany for those forces which remain to cut back on the number of exercises held, the long term trend is not only for proportionately more forces to be based in the UK, but that these forces will have to train here as well. Suitable areas for training in the UK are already in great demand, while environmental concerns have led to pressure to reduce the

> amount of training. There is therefore a real concern that there will be insufficient resources (particularly land) for large numbers of armoured forces to train adequately. Negotiations are already underway to lease training grounds for British armoured forces in Poland, but Britain should also be seriously rethinking the need to maintain what are still comparatively large numbers of heavy armoured forces.

Therefore the continued focus on providing substantial armoured forces (tanks, armoured infantry and artillery) is inappropriate given the likely future threats to British security interests, the inability to move large numbers of such forces to any crisis situation outside Germany, the pressure it places on resources (raising fears of a 'hollow force') and the growing problems in exercising such forces. Even in Bosnia where infantry battalions have found the armoured Warrior infantry fighting vehicle to be extremely useful, offering high levels of protection and firepower, it also requires substantial support (for example in engineers to strengthen bridges to carry its weight), while the deployment of additional Warrior-equipped armoured infantry battalions would not be easy to support. Furthermore, it is difficult to see a major role for main battle tanks and heavy artillery in such operations. Their utility is limited to high-intensity wars, and even then the ability to deploy large numbers outside the UK and Germany, even if such numbers were available, is very limited.

Support arms

Interest in the size of the Army tends to focus on the main combat arms—armour, infantry, artillery and the army air corps. The bulk of the Army's personnel, however, is occupied in supporting these 'teeth' arms. During the 1980s there was a concerted effort to boost the 'teeth to tail ratio' and increase the strength of the combat arms by reducing support functions. In a similar vein the 1994 Defence Costs Study attempted to put the 'front line first' by cutting waste in the support area.[3] Although the Defence Costs Study is to be heartily welcomed in addressing wastage in the armed services—indeed it was probably overdue—the general principle of 'front line first' is more questionable. Again the Gulf War provides an interesting example. 1(BR) Armoured Division's 'teeth' consisted of just two Brigades (and a strong artillery group)—in other words *a quarter of the number of Brigades deployed in Germany* at the end of the Cold War; but the Army deployed 35,000 soldiers in the Gulf, *well over half the size of the British Army in Germany*.[4] In other words, when the Army was actually required to go to war, the tail was greatly increased in proportion to the teeth. A similar phenomenon was encountered in Bosnia, where support for a single infantry battalion was much greater than peacetime levels would suggest. The implication of this is quite clear: if the Army is to offer something more than window dressing in support of Britain's security interests, then the temptation to cut support services in favour of combat arms is to be resisted.

Northern Ireland

A lasting peace in Northern Ireland, if achieved, will have major implications for the British Army. Since the Troubles began in 1969, Northern Ireland has been one of, if not the most important commitment for the Army. Indeed during the 1970s Northern Ireland dominated the Army to such an extent that some senior officers began to think of it in terms of the Army's only major commitment, all else being of secondary concern. Prior to 1969 the normal garrison for Ulster was two infantry battalions and an armoured reconnaissance regiment, totalling approximately 3,000 soldiers. During the 1970s this was dramatically increased, peaking in 1972 at 30,000 Service personnel. The 1995 presence (largely Army) totals 19,000 Service personnel, including 18 Army battalions (or equivalent): six battalions of the Royal Irish Regiment Home Service (formerly the Ulster Defence Regiment); six infantry battalions (accompanied by wives and families) on resident tours of up to two-and-a-half years' duration and six battalions on a six-month unaccompanied deployment as part of the Emergency Tour Plot. Armoured and artillery regiments have been used, converted to 'leg infantry' for this specific role.[5]

Although a rapid reduction in this presence is unlikely—at the very least to allay Unionist fears—equally some reductions will be necessary to accommodate nationalist sentiments. The process of reducing the Army's presence will clearly be driven by a combination of political and security concerns, and a total withdrawal is unlikely in the foreseeable future even if it is considered desirable in the long term. Nevertheless the number of battalions deployed in Northern

Ireland may be reduced, and reduced quite substantially as part of a peace settlement. At the very least the requirement to deploy forces on short tours may be reduced and possibly eliminated. This would help to reduce the overstretch felt by the Army across the board. It would also probably allow a reversal of the 1993 decision not to proceed with the amalgamation of four regiments proposed as part of the 'Options for Change' process. In the longer term, if the resident tours were cut and eventually eliminated, then further savings would be possible.

It is clear that a resolution of the Troubles would be as radical a change for the Army as the end of the Cold War was. Psychologically, Northern Ireland has had probably an even greater effect on the Army than deployment in Germany, while over one third of the Army's infantry battalions (and a steady trickle of armoured and artillery regiments) are preparing for, deployed in, or recovering from Northern Ireland at any one time. Although a settlement is still far from certain, its implications for the Army would clearly be profound.

Reserves and the Territorial Army (TA)

Three types of soldier make up today's army. Regulars are full time, professional soldiers who may be on short-term contracts of a few years, or long-term contracts taking them through to retirement. Regular reservists are former regulars who, for a short period after leaving the service, retain some obligation and may be called up in an emergency. Finally the TA is a volunteer reserve which may be used to provide specialist individuals to supplement Regular units in a crisis, or which may provide whole units of their own to work alongside Regulars. Because they are volunteers they are relatively cheap to maintain, but because they are not full-time professional soldiers (a two-week camp and a number of weekends each year is the usual commitment, though even this may not always be met), their level of training is considerably less. The TA, however, is an essential part of a system with eschews conscription. Without the TA there is no large, ready trained (or even semi-trained) pool of manpower to provide reserves in times of crisis. The TA also allows the full and efficient use of Regulars by enabling them to be released from important but non-operational duties. For example, TA units may take over certain depot or home support duties from Regulars, enabling them to be deployed in a theatre of operations. Finally the

TA can provide specialist skills which the Army might have limited use for in peacetime, but for which a greater demand might emerge in crisis or war. The most obvious example of this is medics, but a less obvious example is cooks: with the increased contracting out of catering the Army is very short of Regular cooks if it deploys in strength. For minimal expenditure therefore (£85.8 million in 1993) the Army obtains a combination of essential specialist skills, force multiplier and general reserve. The role and status of reserve forces are currently under review by the government, and a new Reserve Forces Act is likely in the near future. Three principles should guide the future of the TA. First, that it should be given roles and responsibilities that are realistic given its capabilities and limitations. In particular it should be recognised that the quality and readiness of TA units vary widely. Second, that the TA should be used where appropriate alongside the Regular Army to fill in gaps created by reductions.

The problem of overstretch is widely recognised, and Reservists, including the TA, will therefore have to be used on a more regular basis in any deployment of the Army. To date this has focused largely on the provision of specialist skills and support units. But the day cannot be far off when certain TA combat units are deployed alongside the Regulars as a matter of course in a crisis. This would allow full and proper exploitation of the TA's capabilities, particularly those of the more efficient units. Certain units might have readiness targets which would allow them to be used alongside the Regulars in a crisis deployment. Although they might not be used in every crisis (and a period of intensive work-up training would also be required, meaning that they could not be deployed immediately) nevertheless there would be the expectation that when a significant number of UK troops are deployed (as in the Gulf), TA combat units would be used, particularly as reinforcements or to rotate units through the theatre of operations. Quality could be maintained by use of a cadre system, whereby Regulars provide a framework which is filled out by the TA on mobilisation. Moreover Regular sub-units (company level) might be used within the TA regiment/battalion both to create additional readiness, and to allow deployment on an ad hoc basis when small numbers of additional Regular forces are required to support an existing deployment. The final principle is that the TA continue to provide a reserve which may be used in general war, or which may also be used in support of, or even instead of Regular troops in civil emergencies (much like the US National Guard). This role would allow Regulars to be released from Home Duty for operational duties overseas, and would provide combat reserves in a major conflict, while at the same time recognising that the readiness levels of such TA units might be lower than would be required for a crisis deployment overseas.

Conclusion and recommendations

In an ideal world, the Army should be capable of reacting adequately to all threats to British interests, but financial constraints mean the Army must compromise on this. As a result the Army retains a comparatively large capability to fight a high-intensity war, risking both the underfunding of these same forces and overstretch elsewhere. This chapter suggests that a better balance would be to reduce the current capabilities to fight a high-intensity war. This would enable armoured forces to be adequately resourced (particularly in terms of logistics support) and for a transfer of resources to those other areas which are in severe danger of overstretch, and which are likely to remain in great demand for the rest of this century. Assuming that 1 (BR) Armoured Division deployed in the Gulf can act as a benchmark for the required capability, and adding additional forces for training, rotation and reserves, the following restructuring might be possible :

Table 3.4 Proposed restructuring package

	Granby	Granby+	Current Plans
Armoured Regiments (incl. reconnaissance)	4	6	11
Armoured Infantry battalions	3	5	8
Heavy Artillery	2	2+	3
Field Artillery	3	4	9
Air Defence	1+	2+	4
TOTAL	13+	19+	35

Note: This takes 1 (BR) Division from Operation Granby as a baseline armoured force. Additional units are required for reserves, training, rotation and other contingencies. The proposed package has therefore been termed 'Granby+'. Regular Reserves and the TA would reinforce this in the medium term. For comparative purposes, current government plans are also detailed.

What this demonstrates is that changing the balance away from an unrealistic focus on high-intensity warfare can realise substantial savings, both in terms of manpower and equipment. Additional general purpose infantry and engineer forces—the sort required for low–medium intensity deployments—can therefore be created, the residual armoured elements can be adequately resourced, and some financial savings might still be possible. Major reductions in the support services altering the 'teeth to tail' ratio, however, are unlikely to be cost-effective as the experience of the 1990s has shown: deployment in both the Gulf and Bosnia required considerably greater support than peacetime levels would have suggested. Finally a settlement in Northern Ireland would allow some reductions to be made, perhaps quite considerable. But these have to be balanced against the political requirements of reassuring the people of Northern Ireland, and any residual concerns over security.

4 WEAPONS PROCUREMENT AND THE DEFENCE INDUSTRY

Ron Smith

This chapter examines some of the general questions that any government has to answer in dealing with defence procurement and the defence industry. It assumes that the primary decisions about roles, missions and resources have been made and that the problem is how to provide the weapons needed to meet those needs within the resources available.

As British industry's largest customer, the MoD buys some very high technology equipment. The MoD's procurement decisions therefore inevitably have an effect on British industry and technology. A primary question is whether these buying decisions should be guided by an explicit industrial policy: an articulated view of where the industry should be going. The alternative is to continue the current policy which aims to provide our armed forces with the equipment and services they need at the best value for money.[1] The term value for money will be used repeatedly below and it should be emphasised that best value for money does not mean the cheapest price. Value for money is also determined by performance, reliability, maintenance costs, and a host of other characteristics. Thus there is inevitably a large amount of judgement involved in the determination of best value for money.

Within the general policy, whatever it may be, there are a range of more specific questions to answer.

- What defence industrial base do we require for security of supply?
- Which procurement sources should we use: domestic development, collaborative development, licensed production or importing equipment 'off the shelf'?
- How should contracts be awarded? After competition or to a preferred sole source and what form should they take (fixed price or cost plus)?

- How will defence R&D be directed?
- Should we promote the arms trade, which generated export orders of £6 billion in 1993, and employed an estimated 90,000 people?
- How should we help the arms industry adjust to declining demand?
- What scope is there for reform of the organisation of procurement? As documented in many National Audit Office and Public Accounts Committee reports, the management record of the MoD Procurement Executive is not good. Weapons are repeatedly delivered late, over budget and unable to meet the military requirement. Are these problems inevitable or avoidable?

Answering these questions is inherently difficult. To begin with, defence procurement involves big numbers and a mass of arcane detail. Press and political attention tends to get drawn to small comprehensible wastes, like the cost of decorating mansions of senior officers, rather than the large and complicated wastes such as that resulting from the choice of index to adjust for inflation on the EH101, which will probably cost the MoD almost £100 million. But suppose that the detail is mastered, it is still often the case that the great technological, political, military and economic uncertainties mean that there are no straightforward solutions. But suppose a sensible solution is discovered, getting it implemented in the face of inter-Service rivalry, bureaucratic procedures and industrial obstacles is even more difficult. Even if there is general agreement about basic principles, it is often difficult to apply these principles to particular decisions, where jobs will be at stake, industry and unions will be lobbying very hard and political tempers will be high. Recent examples of controversial procurement and industrial questions include the choice between the Future Large Aircraft and the C130J as a replacement for the RAF's Hercules, a choice the Conservatives finessed, or policy towards the acquisition of VSEL by either BAe or GEC, which was referred to the Monopolies and Mergers Commission.

Background on the defence industry

Procurement decisions are unlikely to be easy, because of the complexity, uncertainty and practical and political difficulties of implementation which surround procurement decisions. The general cuts in defence have meant that spending on equipment has fallen from about £12.5 billion in 1985–86 to £8.3 billion in 1994–95 (at 1993–94 prices). This has prompted substantial restructuring in the defence industry; employment has fallen from 620,000 to about 380,000 and there has been a wave of mergers and acquisitions. However, the restructuring in the UK and Europe has been less rapid than in the US, and further consolidation, perhaps on a European level, seems inevitable. Below the well-known prime-contractors (BAe, GEC, Rolls Royce, VSEL) there is a large pyramid of suppliers to the MoD: 165 contractors received more than £10 million from the MoD in 1993–94; 18 of these received over £100 million. Just over half the total procurement is spent in the aerospace, shipbuilding and electronics industries. England accounted for 90 per cent and the south east of England 35 per cent of the equipment expenditure. The 25 major projects underway at March 1993 had a total value of about £30 billion, of which the largest were Trident (£10.7 billion), the EH101 Merlin helicopter (£3.9 billion), and Eurofighter development (£3.5 billion, production will be at least another £10 billion). Of these projects, 13 showed overspending relative to initial estimates of cost and 9 underspending. Underspends do happen: Trident has cost almost £3 billion less than originally estimated. On average, the projects were just under three years late, relative to the original estimate of their In Service Date.

Particularly since about 1985, there have been major changes in defence procurement policy, emphasising reliance on market forces. Where possible, publicly owned arms production was privatised, and where this was not immediately possible (for example, the Royal Dockyards or the Atomic Weapons Establishment) it was subcontracted to private firms. There was a drive to switch away from cost-plus to competitive contracts where possible (competitive contracts increased from 38 per cent of the total value in 1985–86 to 62 per cent in 1993–94, though the numbers fluctuate from year to year). Prices had to be established before the contract began, payment was linked to performance milestones and more risk was transferred to industry. There was greater willingness to import arms and relatively little concern about acquisition of small- and medium-sized UK defence companies by foreign firms. Although the declared policy was not to have an industrial policy for the defence industry—that was something to be left to the firms—in fact, industrial consequences and domestic industrial lobbying did influence decisions on a number of occasions.

The end of the Cold War and rising costs left the world with massive excess capacity in the arms industry, estimates vary from 30 to 50 per cent overcapacity. The arms industry is steadily restructuring, very rapidly in the US, more slowly in Europe. Although the excess capacity problem is more severe in Europe than in the US, the obstacles to cross-border mergers inhibit rationalisation through acquisition. Instead there has been a proliferation of joint-ventures and strategic alliances. These can either be transatlantic (in aeroengines SNECMA with GE and Deutsche Aerospace with Pratt and Witney) or European (Euromissile, Eurocopter, etc.). The rationalisation of the European defence industry will almost certainly face governments with difficult questions about the foreign acquisition of domestic defence companies. Domestically, any proposals to merge GEC and BAe would also raise difficult industrial questions.

Reliance on competition is also becoming increasingly difficult as concentration within the defence industry means that in many areas there is only one domestic manufacturer and any real competition has to come from abroad.

Industrial policy

While there may be a case for industrial policies to encourage investment and R&D, to improve the skills of British workers and to offset the short-termism of the City, such policies would be more effective if they were broadly targeted at British industry as a whole rather than aimed specifically at defence firms. Formulating a coherent defence industrial policy would be almost impossible because the structure of world industry is changing so fast. If such a policy had any hope of being effective it would have to be formulated at a European rather than a British level. In addition, given their lobbying power targeting the big defence firms risks repeating the mistake of past industrial policy which rapidly changed from 'ministers picking Winners', to 'losers picking ministers', leaving ministers subsidising expensive high technology lame ducks. This is a particular danger with the large defence firms, since their special skill, which is crucial to their survival, is persuading government to give them money. They are very good at this and any measures which would give them an opportunity to mobilise their massive lobbying power would be a dangerous hostage to fortune, which ministers might come to regret if they tied themselves to a particular industrial policy.

Defence ministers must take account of the industrial implications of their procurement choices and these are discussed further below. But in doing this, they should bear in mind that all the great British defence disasters (Nimrod, the Torpedoes, the Harland & Wolff AOR, etc.) started off as well-intentioned attempts to use defence procurement to meet industrial objectives; it is a recipe for disaster to try and hit two targets (military and industrial) with a single shot. Making the decisions on military grounds is hard enough, complicating them with subsidiary economic and industrial objectives, areas on which MoD has relatively little expertise, makes them more difficult. Most of the major procurement decisions are not merely matters for the MoD but are made at Cabinet Committee level, and there may be some role at this level for an explicit analysis of the industrial consequences of procurement decisions, which goes beyond the politics of job losses. However, the primary thrust of policy should continue to be providing the armed forces with the equipment they need at the best value for money; though that this can certainly be done better than it is at present.

The MoD seems prone to repeating certain sorts of mistake, which have been pointed out in a whole sequence of reports.[2] Decisions are made too early, with inadequate knowledge of the technology, leaving the MoD locked into projects which cost more and take longer than expected. There is inadequate risk assessment, particularly with regard to software. Specifications are unnecessarily changed during a project adding substantially to cost. Responsibility is diffused within the MoD so it is difficult to know whom to blame. This is accentuated by turnover in posts which is rapid relative to the life of a large project. The MoD is slow at learning lessons from its failures and applying the lessons to other projects. Thus there is scope to improve the management of defence and obtain better value for money, but it requires changing the defence culture and persuading the Defence ministry, armed forces and industry that military projects are not exempt from the normal laws of economics.

The size of the defence industrial base

Although it is common to refer to the defence industry, defining exactly what it covers is almost impossible. Many of the firms supplying MoD are supplying standard goods, for example petroleum products, and are clearly part of the civilian industrial base. This overlap with the civilian economy is probably increasing, since many electronic components and information systems which in the past would have been custom built for the military are now bought on the open market. Most firms that produce specialised defence equipment are part of diversified conglomerates, though the defence divisions of these firms are often functionally isolated from the commercial divisions.

It is often argued that we need particular parts of the defence industrial base for strategic reasons, to ensure security of supply in time of conflict. In fact, this argument provides little justification for the vast bulk of the defence industry. Currently, the two capabilities which seem to be regarded as essential are nuclear warhead production and cryptography; the things we cannot trust our allies to supply. Everything else is in principle a matter of where we get the best value for money. Often British industry does provide the best value for money, but exceptions have been made, particularly for ships and tanks, which have had special protection for sentimental reasons.

Security of supply is a weak justification because most conflicts Britain is likely to be involved in will (like the Falklands and the Gulf) happen very quickly and be over rapidly. For conflicts of this sort, supply depends on stocks the armed forces hold and amounts our allies are willing to provide (for example, the US supply of Sidewinders during the Falklands war). Particular specialist engineering and electronics skills are important in converting equipment for war; standardisation with allies allows us to use munitions and equipment obtained from them; domestic production capability is largely irrelevant. Industrial lead times are so long, that production for conflict is effectively impossible. Although Britain had a tank industry, no new Challengers could be produced for the Gulf, existing ones had to be cannibalised to get a working unit.

The more important reason to worry about the size of the defence industrial base is to ensure that we have the industrial capability to allow us to acquire weapons tailored to our own needs, in those cases where our needs differ from those of our allies. The MoD distinguishes between capability (the ability to do something) and capacity (how much of it you do). They argue that in the downsizing of the defence industry, a lot of capacity has been lost, but very few capabilities. This seems somewhat implausible, since it is unlikely that the capability to develop new weapons can really be maintained if none are produced. Thus it does seem likely that the capabilities of the British defence industrial base are being steadily eroded. To a large extent this is inevitable, since Britain cannot afford to maintain all these capabilities. What is more worrying is that each of the larger European countries are making similar choices in isolation. Thus there is a potential danger that each will allow a particular capability to erode leaving a US monopoly in crucial areas. But this is a threat that can only be countered within the context of a European security policy: European countries must establish some degree of role specialisation and division of labour in the provision of these capabilities.

Procurement sources

There are four main sources for weapons: the weapons can be developed and produced domestically; they can be developed and produced in collaboration with partners; they can be produced domestically but using technology and designs acquired under licence from abroad; or they can be imported 'off the shelf'. Not all four sources may be available. If a particular sort of equipment is needed very quickly, there is no alternative but to import it. On other occasions neither suitable foreign weapons nor collaborative partners may be available, because other countries do not have the same requirement. However, when all four sources are available, cost tends to decrease as one moves from domestic to imported. This is primarily because the fixed costs of R&D (mainly development) tend to be very large and unit production costs tend to fall rapidly with the length of the production run because of economies of scale and the learning curve. For many major weapons systems neither British governments nor British firms can afford the vast fixed costs of R&D that new systems will require, and existing, let alone future, production runs are too short to amortise these large fixed costs and gain the benefits of learning and scale.

Collaboration helps reduce costs by sharing the R&D between the partners and in principle can provide learning curve and economy of scale benefits in production. In practice, duplication of facilities, differences in requirements, coordination problems, lack of clear control and delays due to different budgetary systems all tend to increase the costs of collaborative projects. Collaborative programmes, such as Tornado and EFA, have suffered a substantial inefficiency penalty, although overall they were probably cheaper than the national alternatives. In many cases even when there has been a common requirement, the difficulties of negotiation between a large numbers of countries has made collaboration impossible. This was the case with the NATO NFR 90 anti-air frigate, where eight countries failed to agree. However, this failure led to an attempt to provide a new model for collaborative production with the Horizon Frigate, an Anglo-French-Italian collaboration, the procurement of which is designed to reduce the cost-penalty traditionally associated with collaboration. Extending European collaboration in research would also bring rewards if organised efficiently. Again this requires developing a more cohesive European defence identity.

Defence contracts

There are longstanding disputes about the relative merits of fixed-price versus cost-plus contracts and competitive tendering versus sole sourcing. The conclusion of the large economic literature on procurement contracting is that there is no unambiguous answer; which is better will depend in quite complicated ways on the circumstances of the particular case. Each form has particular types of disadvantage which have to be traded off, and the trade-off differs. This is why commercial firms use a wide variety of types of contract and procedures.

One disadvantage of competitive tendering is the possibility of 'adverse selection'. The firm that bids lowest is the one that does not understand the difficulty of the contract. This is not merely a theoretical possibility, given the complexity and uncertainty of defence contracts. For instance, one firm which won a competitive contract, Airworks, did expensive damage to the Tornadoes it was repairing. Maintaining the capacity to create competition can also be inefficient when there are economies of scale, so that two small competitors would result in higher costs than a single integrated firm. Against this, monopoly has the disadvantage that it is difficult to evaluate the price and performance of the firm and there may be less incentive for a monopolist to innovate and cut costs.

The disadvantage of cost-plus pricing is that it provides little incentive for the firm to minimise costs. The main advantage of cost-plus is its flexibility—it is easy to make changes to the design after the initial contract has been signed. With fixed-price contracts, the whole contract has to be renegotiated at a time when the firm has great bargaining power. With fixed-price contracts, the firm will also have to add a risk premium to provide an allowance for contingencies. Thus where there are very large uncertainties, particularly during development, non-competitive cost-plus contracts are more common.

Research and development

MoD R&D in 1992–93 was £2.1 billion, down from £2.7 billion in 1991–92. However, most of this is development expenditure related directly to particular weapons; only about a quarter of the total is expenditure on research. Defence research has gone through major structural changes, first with the formation of the Defence Research Agency, and then with the changes following from the *Front Line First* defence cost study.

Whereas traditionally military technology was well in advance of civilian technology, now the reverse is the case in many areas. In the old days defence industrialists used to talk of spin-off (of which there was actually very little) now they talk of spin-in (adapting civil sector technologies to the military). Therefore the technological pay-off to civilian innovation from governments' R&D investments in defence are likely to be very small, but the technological pay-off from transferring civil technologies to defence may be quite large. This in itself requires research and the targeting of the relevant technologies. Since there are a large number of technologies involved, even monitoring and adapting civilian technologies to military needs will become very expensive. Increasingly, defence research will have to be coordinated at a European level if access to a wide range of capabilities is to be maintained.

The research and development will have to be directed by the relative priorities of procurement policy. What numbers of weapons will be required, when, using what technology, to meet which missions, where? It is at the R&D level that the tension between best value for money and lowest price is often most obvious. For instance R&D on reliability and maintainability which is embodied in the weapon tends to raise the purchase price but reduce the total through-life cost of the system; cutting out that work is a false economy. Similarly, technology demonstrator programmes separate from the system itself look like an unnecessary expense, but are usually very good value. In particular, the identification of problems can lead to reconsideration of projects. Many of the big cost overruns have resulted from the MoD becoming locked in: starting full development without investing enough in investigating the technology. The technology which has probably given MoD the most problems in this respect has been computer software production.

Arms trade

Weapons exports are likely to remain a sensitive issue for any government. The government is involved in multilateral negotiations over the regulation of the arms trade, particularly for the EU as a whole, but also in wider bodies such as the successor to COCOM (Coordinating Committees for Multilateral Export Controls). It has to decide how strongly to support DESO (Defence Export Sales Organisation) promotion of British arms. It will have to approve particular sales, some of which (e.g. armoured vehicles to Indonesia) are to countries with poor international records. Relations with other states and foreign policy attitudes are often influenced by arms sales. In particular UK attitudes towards Saudi Arabia have been coloured by the £20 billion Al Yamamah contract.

Probably the largest uncertainty about arms exports is whether they represent an economic benefit. Arms deals are very complicated. The package can involve counter-trade (barter, e.g. Al Yamamah is paid for in oil); offsets (promises by the UK to buy specified amounts from the recipient); technology transfer; linked aid (as with the Pergau dam); credit (the recipient usually borrows the money to pay for the arms from the supplier government, e.g. through the Export Credit Guarantee Department [ECGD], often at subsidised interest rates); insurance (the supplier government usually takes the default risk if the recipient does not pay, e.g. the French government was left with large Iraqi debts). The companies certainly make money from the deals—they would not sell the arms otherwise, but given the implicit subsidy from the supplier government, it is not clear if the national return on the sales (even ignoring the security and human rights consequences) is positive. A Treasury investigation of the real national return on arms exports would be a useful input into the decisions on the arms industry.

Industrial adjustment

Although it is far from certain, given the imponderable security situation after the Cold War, it is quite likely that defence spending and procurement spending will continue to fall. How will this adjustment be achieved? A range of econometric simulations indicate that cuts in military expenditure matched by equivalent increases in other government expenditure have a positive effect on employment, reducing unemployment once the adjustments have taken place. But although total unemployment might fall slightly, the change may be made up of those previously unemployed getting jobs and those previously working in defence factories becoming unemployed. To ensure that these defence workers will get jobs through the market mechanism we need to ensure that the unemployment rate is low. This would have to be achieved through economic policies: a sensible demand management policy, and policies for training, industry and the regions. Between 1985 and 1990 UK military spending dropped from about 5.5 to 4 per cent of GDP; real equipment spending dropped by more than a quarter; and total defence employment dropped by 200,000 (on government figures, which may be an underestimate). Adjustment to this large change was quite smooth and often by 'natural wastage' rather than redundancy because the economy was booming. The later cuts in defence spending, after 1990, which were rather smaller, caused much more hardship and unemployment because the economic recession hit the south, where many of the defence jobs are, particularly hard.

It is often suggested that the difficulties caused by the run-down in the defence industry could be eased by a programme of conversion which would enable the workers to keep their jobs by producing socially useful products. The difficulty with this suggestion is that most evidence indicates that conversion—switching the workers, managers and equipment in specialised defence plants to producing non-defence products—does not work. Norman Augustine of

Martin Marietta was only slightly exaggerating when he said, 'The record of conversion and unrelated diversification is unblemished by success'. Trying to convert arms factories to alternative production is as sensible as trying to convert coal mines to alternative production. In one case the obstacles are geological in the other cultural. Dependence on military contracts appears to induce attitudes and practices, such as performance-maximisation rather than cost-minimisation, which inhibit adaptation to profitable commercial production. When you have put a lot of your life into learning how to do one thing well, it is very difficult to switch to doing something completely different. Such a switch is hard for individuals, though they can manage it; it is almost impossible for whole plants or organisations. Keeping the team together means that procedures, expectations and social interactions reinforce the old attitudes and behaviour patterns. The few arms firms that have managed to diversify into unrelated areas (e.g. Racal into Vodaphone or Matra into cars) did it by setting up new plants, not by converting their existing defence plants.

Defence workers and managers have great skills, but they have to be moved out of their military teams so that these skills can be used elsewhere in a civilian environment. How can this redeployment be achieved? Regional policies, to revitalise areas characterised by high structural unemployment, will be an important aid to redeploying defence workers. Although central government support will be important, these should be driven by local initiatives which can take advantage of the particular skills, strengths and opportunities of the area. This local orientation will be crucial for defence dependent areas, because defence plants are so heterogeneous: no one model will work for all areas. Although they are all involved in military production, facilities in aerospace, shipbuilding and dockyards, electronics, ordnance and armoured vehicle production, are very different in types of skills employed, land-use, concentration, etc. Having a single policy for all defence-dependent areas is unlikely to work. Encouraging local enterprise initiatives is the preferred option.

Procurement reform

Although there is undoubtedly much fraud, waste and mismanagement in the Ministry of Defence, it is much easier to identify *ex post*, when it has been identified and documented in NAO reports. At the time they are happening, the mistakes can be remarkably difficult to detect and prevent. In addition, the large cuts following 'Options for Change' and *Front Line First* have had an impact. Therefore while there is scope for some efficiency gains, these are unlikely to be very large. Nor does international experience suggest that there are any magic formulae to improve defence procurement; buying weapons will always be a difficult and costly job, dependent on the skills of the managers rather than the detail of the procedures imposed on them.

The one area where reform seems most needed and change seems inevitable is the development of more European cooperation in procurement, R&D and the restructuring of the arms industry. This is obviously not a matter for the UK alone, it has to be negotiated with other countries who have very different procurement styles and very different relations with their arms industries, which are often state owned. Nor can development of this dimension be confined to procurement. Procurement initiatives will only work in the context of the development of a wider European defence and security identity. The economic pressures for cooperation are obvious and are making national independence in defence unaffordable and national arms industries non-viable. The alternatives to European cooperation—spending vast sums to attain minimum efficient scale of production, dependence on the US or acceptance of effective neutrality—are likely to appear unattractive to the larger European states. The major imperative for the next British government is to overcome the obstacles to effective European cooperation.

5 RETHINKING BRITISH ARMS EXPORT POLICY

Susan Willett

The Matrix Churchill scandal highlighted the British role in the international arms trade in 1992. Lord Scott's subsequent inquiry into the affair revealed a less than perfect system of export licensing and exposed the role of government ministers in circumventing government guidelines on arms transfers.[1] The evident gap between practice and policy which the inquiry exposed suggests that a re-evaluation of UK arms trade policies is long overdue.

The Matrix Churchill affair was the first in a catalogue of press disclosures about malpractice in the UK's arms trade. Revelations included the use of British aid for the Pergau dam in Malaysia as a 'sweetener' to secure large defence contracts;[2] the controversial sale of BAe's Hawk 200 combat aircraft to the Indonesian regime, which have been used in bombing raids against defenceless civilians in East Timor; allegations that British companies supplied small arms and ammunition to Iran and the former Yugoslavia in breach of sanctions;[3] and accusations of corruption and bribery concerning the Al Yamamah deal, the largest arms sale in the UK's history, signed with Saudi Arabia.[4]

Prior to such revelations the trade in conventional weapons and military related technologies had been a relatively uncontroversial issue in the UK. A general consensus prevailed, shared by the leading political parties and trade unions alike, that arms transfers were good for the British economy as they contribute to the balance of trade, industrial output and employment. Moreover, arms sales are viewed as a useful foreign policy lever, maintaining and extending the UK's sphere of influence. And there was a general belief that government responsibility in administering arms transactions ensured that British arms were only delivered to reliable client states whose security interests were strongly associated to those of Britain and its allies. However, recent government practices question the validity of these assumptions. A wide spectrum of the British public now fear that UK arms trade is rife with abuse, malpractice and corruption, and that the government has irresponsibly promoted the proliferation of conventional weapons. Such serious indictments demand investigation—not least to clarify what is true or false—but also to rectify the loopholes in what appears to be an extremely porous arms trade policy.

The Scott Inquiry in 1994–95 and the hearings on weapons proliferation by the House of Commons Select Committee on Foreign Affairs promoted the first public debate about the UK's conventional arms trade policy, raising some critical questions about the benefits of the arms trade, the conduct of government, the accountability, transparency and adequacy of existing arms trade control mechanisms. This chapter attempts to set out some of the issues that must be addressed if Britain is to reassert a responsible arms trade policy.

Existing export controls

As a major military power with extensive international interests the UK long ago established a national system of management of arms exports through licensing and has restricted or prevented certain exports in certain circumstances. However, the existing system of controls is based on a set of loosely defined guidelines rather than a clearly codified set of rules. This allows considerable flexibility in the system but also leaves the process open to abuse.

Arms trade management, whether unilateral or multilateral, implies rules for reporting or licensing sales in specific situations, as well as in market sharing arrangements. Restrictions of sales involve the banning of specific transfers or the implementation of more general embargoes, either to specific countries or through specific types of military hardware. The reduction of transfers involves more sweeping measures, such as qualitative and quantitative limits or ceilings and conscious government policies or international agreements designed to limit specific technology transfers. In the process of formulating restraints, there are often trade-offs between restrictions on the transfer of weapon systems versus controls on production technology or know-how which enable the fabrication of military hardware. Export controls and licensing procedures are implemented at an institutional level by the various bureaucratic actors which have a stake in the arms export process.

The Department of Trade and Industry (DTI) is authorised to apply export control legislation and issue export licences. This is carried out by its Export Control Organisation (ECO). Before approving a licence, the ECO has to obtain clearance from other government departments, such as the Ministry of Defence and the Foreign Office, and in the case of nuclear materials, the Department of Energy. The MoD reviews proposed sales in the context of security implications, while the FCO advises on potential geo-strategic considerations. Where conflicting advice arises between different departments the matter is normally settled at either interdepartmental level or ultimately through Cabinet Committee.

Qualitative controls on weapons transfers, including the modification and downgrading or the outright banning of weapon system transfers, are largely the outcome of military concerns about the release of sensitive technologies. Indeed such considerations are supposed to be the principal export licensing criterion in the UK, lending great weight to the MoD's role in the export control process. In addition to the effect on Britain's security and geo-strategic concerns, other criteria and factors are supposed to influence the arms control equation, including questions of regional stability, human rights and internal repression, proliferation issues, and economic interests. However, the government applies a great deal of discretion in exercising such criteria.

Furthermore, UK guidelines assert in principle that arms should not be transferred to regions at war or to dictatorial regimes likely to use weapon systems for internal repression. Yet a significant number of export licences which are requested and granted go to regimes such as Indonesia and Nigeria known for their ruthless internal policies and human rights violations. At times the government flagrantly ignored embargoes or international control regimes, by allowing arms sales to embargoed countries such as Iraq, Iran and South Africa.

End-user certificates exist to prevent the retransfer of sensitive military technology from one recipient to another. But the enforcement of end-user certificates is thought to jeopardise market access by identifying the supplier as unreliable. For this reason, the UK imposes end-user certificates only in specific cases where danger of re-export is considered great. The government relies on intelligence information to track down harmful re-exports and the threat of future sales bans to discourage them, but clearly in many cases such provisos have broken down or intelligence information has been blatantly ignored, as in the case of the Matrix Churchill affair.

In general the UK adopts a case-by-case approach to arms transfers. Both the MoD and the FCO have tables which rank weapon systems in order of technological sensitivity and rank potential customers according to security risk or political acceptability. These lists are then compared as a rough guide to ascertain suitable customers. The Treasury's ratings of creditworthiness is a further, if secondary, criterion. There are, however, many contradictions within this system of licensing and control, the most blatant incongruity being between the Ministry of Defence's regulatory function and its role as an arms export promoter through its Defence Export Sales Organisation (DESO).

The lack of accountability and transparency

Under the present system, the British parliament plays no role in decisions about arms sales. The accountability of ministers to parliament is the chief check on the abuse of power; however, in practice a government with a large enough majority—as with the Conservatives throughout the 1980s—is rarely accountable in any stringent sense. This is particularly so where there is no statutory requirement for consulting or informing parliament about potential arms sales. In general if an MP wishes to raise questions about arms exports they have to be included in debates on foreign and defence policies and more often than not end up relegated to the bottom of the agenda where larger policy issues are at stake. With no established Bill of Rights, or Freedom of Information Act and/or strong tradition of judicial reviews of legislation or government actions there is little capability for monitoring government accountability in such a sensitive area as weapons transfers. Revelations about malpractice rely upon individual disclosures. The Pergau dam affair only came to light because a senior civil servant took it upon himself, in an act of conscience, to reveal the misuse of aid. *The Economist* has observed that

> The details...are embarrassing enough, but the real scandal is that there is barely any scrutiny of the export of arms, or the equipment used to make them, by anyone except government ministers who approve, or promote, the trade. Arms are not like any other export product. To whom a country sells them is one of the most significant, and controversial, foreign policy issues facing any government.[5]

The lack of parliamentary accountability and the shroud of secrecy which surrounds arms sales in the UK has been justified on the basis of commercial confidentiality and respect for the national security of the recipient nations. However, not all major suppliers impose such restrictions. In the US and Germany, for example, government accountability is strictly adhered to and all major sales are debated in the various legislative bodies. Arguably the secrecy which surrounds UK arms sales encourages abuse. As *The Economist* contends

> the idea that arms sales should never be scrutinised by anyone, or that arms-export policies should never be publicly debated is preposterous. At the very least an all-party committee of Privy Counsellors...should review arms deals and report to parliament when the government's own stated policies are breached....Nothing justifies leaving government ministers to operate in the shadow, answerable only to themselves.[6]

To date governments have defended the present closed system in order to avoid scrutiny which would undoubtedly affect the level of UK arms sales. In a highly competitive global market environment in which the goal of arms export maximisation has become a major priority of UK trade policy, the resistance to greater transparency and accountability in arms trade dealings has become an institutionalised norm, maintained and reinforced by those with a powerful vested interest in the arms business.

Maximisation of arms sales

In rationalising its continued promotion of arms sales, the UK government has consistently fallen back on the theme that cutting its 5 per cent share of global arms exports without multilateral restraint would have no significant beneficial effects, and if the UK doesn't supply arms then someone else most certainly will. The most influential factor contributing to the government's active promotion of arms sales is the relatively small domestic procurement base which results in high unit costs for weapon systems and high military R&D costs. Eventually rising unit costs and budgetary constraints force a choice between dependence on imported technologies, specialisation in weapon systems of particular national importance, the abandonment of defence industrial and technological capabilities or defence exports. By promoting arms sales, greater economies of scale are generated, R&D costs are spread and companies are able to move up the learning curve, which helps reduce domestic procurement costs.

Over the last decade the desire to maintain a strong defence industrial base within the context of diminishing resources has left the UK government with little option but to maximise exports. Huge resources have been mobilised to this end. At the highest political level ministers have become active in arms export promotion and support institutions were charged with the mission of optimising arms sales. Considerable resources have been allocated to promote British arms transfers, mainly through the Defence Export Services Organisation (DESO), a department within the Ministry of Defence, first set up by the Labour Party in the 1960s.

With over 700 staff and a budget of £21.2 million, DESO helps companies to obtain licences by coordinating inter-ministerial discussions of sales; it promotes British defence products by organising yearly exhibitions such as the Army and Navy Exhibition held at Aldershot in 1993; it produces the Defence Equipment Catalogue; it engages the support of the armed services and defence attaches overseas to help promote sales, and provides advice and market surveys. It is interesting to note that no other manufacturing sector receives such generous support from the government. It is worth speculating what the economic returns might be from civilian sectors if they received a similar level of government subsidy and export promotion.

The government has also paved the way for arms export maximisation by extending credit cover on major deals. In June 1988 an extra £1 billion of public money was made available to the Export Credit Guarantee Department specifically to boost large overseas defence sales.[7] The ECGD insures UK exporters against payment defaults from high risk countries.

In maximising arms sales many of the general guidelines for restricting arms transfers to countries or regions, on the basis of human rights violations, internal repression and wars, were overlooked or selectively applied. For example the ban on the sale of 'lethal weapons' to Iran and Iraq was conveniently ignored enabling the approval of export licences for a whole range of equipment destined for military or military related purposes in both countries. End-user requirements became less specific as the government relaxed its procedures in pursuit of sales.

The general atmosphere of laxity towards arms trade controls in the 1980s enabled suppliers to use private arms dealers, obscure shipping lanes, middlemen and false end-user certificates to disguise real destinations in their attempt to secure sales. Often such deals took place with the tacit approval of government or even its active involvement.[8] The government also turned a blind eye towards questionable 'dual-use' trade. For example the authorisation of the sale of a Marconi surveillance radar system to South Africa, in 1983, in clear contravention of the UN arms embargo, was justified on the basis of its civil application, yet the radar ended up at a military base in the Eastern Transvaal.[9] The supply of equipment to Iraq by Matrix Churchill and Sheffield Forgemasters reveal a general failure to screen recipients, apply and monitor end-user controls or exact compliance.

These cases illustrate the fact that the existing export guidelines and control procedures are only as effective as the government of the day intends them to be. During the 1980s the guidelines were submitted to manipulation and misuse by a government determined to sell weapons at all cost. The political determination to implement controls has to exist for controls to be operative. Clearly, in the 1980s this was not the case.

The confidence with which the government was able consistently to circumvent or ignore its own guidelines and established practices can be seen as a function of the impunity guaranteed by the general lack of accountability and transparency in arms export procedures. Through such misuse of the system the government did achieve its objective of increasing the UK's share of the global arms market during the 1980s. This success, however, appears to have been of short duration given the dramatic decline in global demand for weapon systems. In the meantime the country's reputation as a responsible arms trader has been damaged.

The hidden cost of the arms trade

The major justification for maximising UK arms sales have been their alleged economic benefits. The consequential costs of adopting a policy of relatively unrestrained arms exports—including short-term budgetary costs and long-term political and economic consequences—are rarely scrutinised in official circles. The assumptions that arms exports help the balance of payments, maintain employment, and enhance the nation's industrial skills and technology base have been seriously challenged when subjected to critical academic analysis.[10]

Arms exports generate roughly 100,000 jobs representing some 2 per cent of manufacturing employment. As a percentage of total exports arms sales fluctuate according to demand but have averaged around 1.8 per cent during the last decade.[11] In 1993 the official MoD figure for arms deliveries was £1.3 billion. While significant, the benefits of these arms exports are less meaningful when the value of government subsidies are removed from annual totals. Total government subsidies for arms export promotion were estimated at £384 million in 1994, representing roughly one-fifth of the total income generated by arms exports.[12]

Arms exports are subsidised by the government in a number of ways; through financing the development of the weapon system:[13] by paying for the promotion and marketing of weapons through the MoD's Defence Export Sales Organisation;[14] providing export credits and insurance through organisations like Eximbank and the Export Credit Guarantee Department (ECGD), which make cheap loans available to the buyer, and give compensation to the seller if the buyer defaults on payment, a not unusual occurrence; by giving the recipient aid to finance part of the purchase;[15] and by making political and economic concessions to the buyer to win a deal.[16]

Moreover, real returns from arms exports are far less than official statistics lead one to believe. Smith states, 'The financial complexity of the arms export process not only makes it difficult to calculate the real return, but provides a variety of routes for "invisible" cross-subsidization, disguising transfers which would be unacceptable if they were apparent.'[17] The real returns on arms sales are more often than not obscured by 'offsets'.

A number of sources indicate that for most sales outside of the Middle East, most notably in the Asian Pacific, a 100 per cent offset target has become standard practice in arms deals.[18] Essentially offsets can be direct or indirect. Direct offsets comprise the participation of the recipient nation's industry in some aspect of the contract for supplying foreign defence equipment. For instance, the acquisition of a tank might involve final assembly in the purchasing country together with subsequent repairs and servicing.[19] Indirect offsets involve goods and services unrelated to the specific foreign defence equipment being purchased. Examples include the supplying nation's government or contractor agreeing to purchase some other item of defence equipment or civil goods from the recipient nation, for example Tornadoes for oil. Such an item then has to be marketed separately. In some cases the resale value of goods may be substantially less than the notional value in the contract. In addition, 'offset purchases' from the recipient country may require accepting poor-value products, transferring technology, or investing in established industries in the buying country, which may be charged against the aid budget.

Offsets can be viewed as both a marketing tool and the means of financing weapons sales, but they are not greatly liked by industry. In an increasingly competitive market contractors have been willing to go along with these arrangements to secure sales, but such arrangements often run counter to their long-term business interests. While prime contractors may benefit from immediate sales, subcontractors often lose their business to co-producers in the recipient nation. In this manner the supplier nation exacts an industrial and employment cost. For instance the US General Accounting Office has calculated that although the Korean Fighter Program will result in more US jobs gained than lost, the balance is very delicate. And in the long run the fear is that the transfer of US aerospace technology is likely to result in the creation of a formidable competitor in the future.

A US survey of major arms sales during the period 1980–87 by the Commerce Department's Bureau of Economic Analysis identified $35 billion in military exports involving offset arrangements. The reported value of the offset deals was an estimated $20 billion, or more than 57 per cent of the value of the original transfers.[20] So far there has been no similar enquiry in Britain into the scale of offsets commitments or their opportunity costs. But UK firms are increasingly involved in offset arrangements which suggests that far less of the reported economic advantages in arms sales actually accrues to the UK than existing arms sales data lead one to believe.

The National Audit Office has noted that there are several risks involved in offset arrangements for both supplier and recipient nations.[21] When acquiring foreign defence equipment, offsets are a means by which the purchasing government obtains compensating work for its national industries and products. Financially and politically these deals can look very attractive to the recipient nation. However, it is always hard to evaluate whether or not the purchaser's national industry would have received the work anyway. And such agreements are difficult to control and monitor.

In the US there is growing concern that 'offsets are economically inefficient and that they foster foreign competition with US manufacturers....Cynics might also add that they appear to be a vehicle for pork-barrel politics and petty corruption on a grand scale.'[22] Since offsets frequently involve taking business from UK firms and giving it to foreign suppliers, they ultimately diminish the net economic benefits of arms sales to the UK economy.

The National Audit Office has drawn attention to certain problems which arise with credit financing arrangements

> in assessing alternative options, the full costs and benefits of different financing arrangements are not fully addressed. If this is not done there is a risk that what might appear to be a good deal is, owing to credit finance arrangements, not actually the best option.[23]

For instance problems may arise as a result of movements in the exchange rate affecting the ability of the purchasing nation to repay 'loans'. Another danger, illustrated by the case of Iraq, and Iran before that, is that a country which is classified as an ally one moment can rapidly turn into the 'next Hitler', thus leaving loans unpaid. The cumulative costs of writing off bad foreign loans for military sales is ultimately borne by the taxpayer.

On closer examination it appears that the true contribution of arms exports to the UK's macro-economic performance is at best marginal. It is rather at the micro-economic level that the economic significance of arms exports reveal themselves. The top UK defence companies represent the nation's high technology capabilities, and defence exports account for a significant part of these companies' turnover. These high-technology sectors are also the most critical for maintaining high-skilled employment.

Individual defence firms make significant profits from arms exports

Table 5.1 Dependence on exports of major UK defence contractors mid-1980s

Firm	Global Rank	Products	Arms production as a % of total output	Arms exports as a % of total exports
BAe	7	Aircraft/missiles	54	55
GEC	14	Electronics	5	45
Rolls Royce	39	Aeroengines	40	42
Thorn-EMI	41	Electronics	20	35
Ferranti	44	Electronics	80	40
VSEL	55	Ships	100	30

Source: K. Krause 1992

but these profits derive from the taxpayer rather than export clients.[24] In the long run, however, the subsidies for arms exports to individual companies may help to undermine the very survival of the companies in question, as an increasing percentage of military technology transfers take the form of production know-how, which inadvertently helps to foster competitors in the market place.

In a buyers' market the pressure to enter into licensed production, direct offsets and joint ventures is considerable. Essentially such arrangements represent the transfer of production know-how and have long-term implications for proliferation, competition and security. Unrestrained arms transfers are linked to conventional weapon systems proliferation and the growth of aggressive regional powers that may pose a potential threat to the UK in the long run.

The most notable security implication of co-production deals is the irrevocable transfer of defence industrial technologies and manufacturing know-how not only in conventional weapons but also for the possible development of long-range missiles and weapons of mass destruction. In addition, co-production deals could have the effect of fermenting regional militarism through the proliferation of conventional weapons, while eroding suppliers' control over transferred military capabilities. They foster a more competitive arms market which will lead to less discriminate sales and still more technology transfer in the future. The expanding numbers of arms suppliers profoundly complicates efforts to achieve limits on international arms transfers through negotiated arms control or multilateral arms embargoes.

The most costly effect is when friend turns to foe and poses a direct security threat. This has happened twice to the UK in recent history, namely in Iran when the Shah was overthrown by the fundamentalists, and in Iraq when Saddam Hussein provoked the Gulf War. In the latter case, the UK involvement cost an estimated £2.5 billion.[25] Once contributions were made from allies the bill for the taxpayer was greatly reduced. But the point here is that the total cost to the allies of containing the effect of proliferation was in excess of $80 billion. All the major suppliers, including the UK, were implicated in supplying the weapons and technologies which enabled Saddam Hussein to build up the fourth largest army in the world. In the 1980s the Iraqi arms market was seen by many, including the UK government, as a bonanza. But in the end it was the suppliers who had to pay the price of proliferation.

Without detailed data, assessment of the economic costs of the arms trade are speculative. However, even on the limited evidence which exists it can be asserted with relative confidence that at best UK arms transfers provide marginal short-term gains for the UK economy. Once all factors are taken into account, the opportunity costs for the UK and the international economy of promoting arms sales result in a long-term net cost to the UK economy.

If the economic advantages of arms sales are more illusion than reality, why then does the state continue to promote arms exports? There are very clear substantial vested interests in the defence industrial base which have over time developed a powerful lobbying force able to exert considerable influence over government decisions both on procurement and trade. For instance, emerging evidence suggests that in the 1988 Malaysian deal, the British government agreed to a secret pact to subsidise exports for British Aerospace and GEC, both companies having curried favoured status with the government.

Policy recommendations

There is a clear need for a major review of UK arms transfer policies. Such a review would coincide with a number of national and international initiatives designed to improve arms transfer controls in the post-Cold War era. In the section below a number of recommendations are listed aimed at improving UK policy.

Transparency and accountability

There is a need for more public notice of arms transfers and destinations, and a stricter criteria about the supply of arms to countries engaged in proliferation or arms build-ups. Although Britain is a highly developed democracy the tradition of secrecy and governmental privilege which surround defence matters plus the vested interests of the 'Military Industrial Complex', means there is little public accountability in the decision-making process over arms transfers. Parliament plays no formal role in the decision-making process, and there is no statutory requirement for it to be either consulted or informed about current or future sales. While both the Public Accounts Committee and the House of Commons Defence Committee monitor arms sales and related issues, their influence over this aspect of government policy is severely limited by their restricted access to information.

Recent disclosures suggest a clear case for greater transparency and accountability in the UK's arms trade policy. At the very least there should be an all-party committee which should regularly review arms deals and report to parliament when government regulations are transgressed.

While the UK government has to be commended as a prime mover behind the creation of the UN Register on Conventional Weapons it should be encouraged to support moves towards prior notification of arms sales and the register of all production capabilities.

An arms sales policy audit

Using the full investigatory and auditing powers at its disposal the government should conduct a comprehensive arms sales policy audit that reviews the full costs of the UK government's efforts to promote arms sales, and takes account of the potential costs to UK security of continuing the audit of unrestrained conventional weapons exports. The findings of this policy should be used as a guide for reforming and improving UK arms transfer policies and practices.

Export controls

The results of the UK government's relatively unrestrained approach to arms sales and the laxity with which arms export guidelines have been interpreted suggest the need for a more restrictive domestic arms export policy and greater efforts in establishing multilateral export controls for conventional weapons. End-user certificates should be issued more often and arms transfers should be effectively monitored. Compliance should be enforced through the threat of substantial fines or sanctions.

At an international level the government should be encouraged to re-inaugurate the P5 or the P4 talks (if China will not re-engage) on arms transfer restraint and regional arms control initiatives. New types of weapons should be restricted to regions where they are not already present. Equally the government should be persuaded to support the moves towards greater harmonisation of European arms export controls, endorsing the moves towards greater restriction and control as advocated by the Germans, rather than holding out for the more *laissez-faire* policies at present advocated by both the British and French governments.

Offsets arrangements

The DTI should be required to notify parliament of co-production or licensed production of major weapon systems or components with any country, and to explain why it is in the interest of national security to provide the country with an independent arms capability. The report should list the articles to be produced, how many and by whom, restrictions on third-party transfers, and a description of controls incorporated into the agreement to ensure compliance with this. The report should be made available to the public so that co-production trends can be monitored by independent analysts. In addition an 'impact statement' should be required for each co-production deal, examining the consequences for regional security, trade and employment.

The government should develop a set of guidelines on co-production to ensure compliance with co-production agreements. Parliament should be notified of the details of all co-production MOUs (Memoranda of Understanding). Notifications or reports should include a section on whether compliance-related provisions have been made.

Code of conduct on arms exports

The government should support the proposals for a British and European Code of Conduct on Arms Exports. The premise of such legislation is that states which are not democratically elected, that abuse human rights, or are at war, are inherently less stable. And states which do not share limited information on weapons procurement with the UN Register on Conventional Weapons are innately untrustworthy; therefore such states should be denied arms transfers. As a general guideline in international negotiations on multilateral conventional arms control the UK government should be encouraged to organise around the highest common factor such as the German standards, and not the watered down level of minimum controls which the UK government has been emphasising in recent debates with international allies.

Conclusion

On the issue of arms restraint the British government has consistently expressed a preference for multinational approaches to the control of arms transfers in the belief that the British share of the global arms market is too small for a unilateral control effort to have much effect. To its credit, in the promotion of multinational approaches, the UK played a prominent role in establishing the UN Arms Trade Register which was formally instituted in December 1991. But as Spears has observed, British enthusiasm for greater international transparency in arms sales sits uneasily with Britain's unwillingness at the domestic level to reveal details of arms transfer agreements.[26] Spears is also cynical about the possibility of a change in UK arms trade policy in the near future. She anticipates a continuation of Britain's 'ambivalent' approach of appearing concerned about the global increase in conventional arms spending while maintaining a policy of aggressively marketing British wares in the global arms bazaar.[27]

In light of revelations about widespread malpractice in the UK arms trade there is a clear case for urgent reform in the existing arms trade policy. A set of criteria needs to be adopted which has greater powers of enforcement in managing the arms trade system than exist through the present system of guidelines. To complement and reinforce such procedures there is an urgent need for greater public scrutiny of arms transfers and destinations, and stricter rules about the supply of arms to countries engaged in proliferation, arms build-ups and conflict. Adopting a responsible arms trade policy, however, will not be achieved by employing a single measure for all situations. A clear assessment needs to be made of the policies and intentions of all potential recipients.

Adopting a more responsible arms transfer policy, however, will depend on a reduced government and economic stake in military production and therefore on a redefinition of national security which downplays self-sufficiency in weapons production. For far too long the vested interests of the Military Industrial Complex have set the terms of the debate and been allowed to bend the rules in arms trade practices, to the detriment of the UK's international standing and often in flagrant violation of the principles of international law.

If the UK is to reassert itself with dignity and authority then the contradictions embodied within its domestic arms trade system must be eliminated which demands a major overhaul of existing practices and structures, particularly in the areas of public accountability and transparency. If on the other hand parliament is denied a formal role in the decision-making process, and there is no statutory requirement for it to be either consulted or informed about current or future sales, the misuse of power and the spread of corruption in the UK arms trade are likely to continue.

PART 3
BRITAIN IN EUROPE

6 BRITAIN AND EUROPE'S COMMON FOREIGN AND SECURITY POLICY

Trevor Taylor

The characteristics of the CFSP commitment

Two elements are of particular note in Title V of the 1991 Maastricht Treaty, which commits the EU to develop a European Common Foreign and Security Policy. The first is the scale of ambition that the Treaty expressed: EC members would no longer simply consult but would cooperate to develop comprehensive common policy.[1] The Treaty thus reflects an apparent belief on the part of member governments that their major interests are harmonious and will be pursued collectively rather than individually in order to achieve success. The second is that the Treaty abandons any effort to quarantine 'foreign policy' from 'security' or even 'defence policy'. In the words of the Treaty: 'The common foreign and security policy shall include all questions relating to the security of the Union, including the eventual framing of a common defence policy, which might lead in time to a common defence.'[2] The Treaty rejects the notion that the promotion of European integration could remain a purely civil activity. The second paragraph of the 1987 WEU Platform document was thus reinforced: 'We recall our commitment to build a European union in accordance with the Single European Act....we are convinced that the construction of an integrated Europe will remain incomplete as long as it does not include security and defence.'[3]

These changes did not reflect a sudden change of direction in 1991, but instead reflected several years of experience and practice, with European Political Cooperation (EPC), involving both regular meetings of foreign ministers and officials at many levels, as well as enhanced consultation on a day-to-day basis using the Coreu communication system. Having begun in the early 1970s, EPC stemmed from the need to relate the EC's external economic policies, under the Commission, to wider political matters. The end of the

Cold War, offering both challenges and opportunities, exposed the need to take EPC further, although during the 1980s states of the EC were often drawn to give their grouping 'a political profile equal to its growing economic importance'.[4]

This chapter first examines the conceptual and empirical background behind both the assumption of common European interests and the recognition that foreign policy cannot be insulated from security and defence, then offers some areas where Britain might make specific contributions to the CFSP.

CFSP and assumptions about shared interests

Initially it is helpful to consider the ideas of national interest and shared interests which underpin the theoretical feasibility of a CFSP. At issue is the extent to which the states of the EU feel that their interests are harmonious, even shared, and how far shared interests can be expected to emerge easily.

Perceptions of national interest are invariably a matter of judgement. This is a true whether one adopts a pluralist or a realist view of foreign policy. In the pluralist perspective, the stress is on the fact that governments rule societies which themselves include different interest groups: if Saudi Arabia has the resources to spend £200 million on either more British Tornadoes or on British Westland helicopters, the populations of Somerset and north Lancashire will likely have a different view of which choice most serves the British national interest. Similarly if the British government decides to put less emphasis on defence capabilities and more on foreign aid in UK external policy, different groups in the UK will gain and suffer. On the other hand, the realist perspective ignores differences within countries and stresses that governments should simply seek to maximise the power of their state. However, the optimum choice even in this model must remain a matter of insight and judgement—the British and French governments of 1956 thought they were pursuing faithfully their national interests when they invaded Egypt. Even the preservation of the state does not automatically justify the payment of any price: the Czechoslovakian government accepted the takeover of its territory by Germany in 1938 and 1939. Arguably this was in the interest of the Czechoslovakian people which did not suffer casualties in fighting where they would anyway have lost. The tourists who flock today to Prague do so because this is a wonderful old city not spoiled by war. Thus even what Lawrence Freedman has asserted as 'vital interests—those bound up with the very continuity of the state'[5]—are a matter of judgement rather than objective fact.

Yet governments constantly assert that there is such a thing as the national interest and indeed the national interest can be conceived as that which yields the optimum benefit for the welfare of a population as a whole. Bureaucratically, governments need to specify their definition of the national interest in order to give coherence to the actions of the large, variegated bureaucracies involved in foreign affairs. But the national interest must remain a contested concept, whose character is based on judgement. Thus the feasibility of a CFSP depends on the extent to which the members of the EU can agree on common judgements about their interests, given their different historical backgrounds, geographic location, political systems, economic structures, and sources of information.

It is helpful at this stage to view the concept of a shared interest at three different levels, and to forecast that agreement will become harder with movement towards the third level. The first level of interest comprises the broad characteristics of a desired world: clearly a CFSP cannot be viewed as feasible unless all the EU states are believed to have a common general view of the sort of world which they see as suiting their interests. Significantly the great majority of political figures in the EU favour a world where liberal democracies and market economies are dominant, where international disputes are not resolved by the threat and use of force, where the United Nations and other international bodies constrain possible anti-social behaviour by governments, and where international law is strong. Paragraph 2 of Article J.1 of the Maastricht Treaty spelled out the sort of world which the CFSP is to promote.

The objectives of the common foreign and security policy shall be:

- to safeguard the common values, fundamental interests and independence of the Union
- to strengthen the security of the Union and its Member States in all ways
- to preserve peace and strengthen international security, in accordance with the principles of the United Nations Charter as well as the principles of the Helsinki Final Act and the objectives of the Paris Charter

- to promote international cooperation
- to develop and consolidate democracy and the rule of law, and respect for human rights and fundamental freedoms

The second level of shared interest concerns the desired outcome in a particular situation. A commitment to a CFSP implies that the EU states' overall idea of a desired world can regularly be applied in specific situations so that there can be agreement also on the goals of policy in a particular situation—for instance, to liberate Kuwait, to halt slaughter in Rwanda, or to prevent further nuclear proliferation. Thus the EU has adopted numerous statements on behalf of the Presidency and 'common positions' on subjects such as the embargo against Serbia and Montenegro.[6]

The third level of interest is the most demanding and is often troublesome even within states. A shared interest here requires, not just agreement on a policy goal, but also on the best means to achieve it, the prices to be paid, and the risks to be run. This level of shared interest was not achieved in the Kuwait crisis by the EC states since not all states judged that it was worth risking the lives of their forces in the liberation of Kuwait, although all EU states saw the use of force against Iraq as justifiable. The difference between the second and third levels corresponds roughly to the Maastricht distinction between 'common positions' under Maastricht Article J.2 and 'common action' under Maastricht J.3.

The EU states have managed some 'joint actions', notably the promotion of the European Stability Pact, support for humanitarian aid convoys in Bosnia, the commitment to administer Mostar for two years, and backing for an indefinite extension of the Nuclear Non-Proliferation Treaty.[7] Moreover, in the realm of external policy in general, rather than CFSP in particular, the member states have long shown a capability to act together, most obviously by entrusting many trade and aid issues to the Commission on a day-to-day basis, and by establishing the EC as an international actor with a legal personality able to enter into international commitments. The EU has 100 diplomatic missions around the world, is an observer at the UN and other international organisations, and has 150 missions accredited to it.[8] Yet, apart from economics, those most optimistic about a CFSP must recognise that agreement on the third level of interest, generating common action, will often be hard, perhaps impossible. The traditions of the four members of the EU which were neutral during the Cold War, the different levels of economic resource available to individual members, the varying readiness to accept military risks among member governments, and the variations in size, location and history of EU members all indicate the difficulties. Britain and France are former great powers accustomed to having a significant say in the broad development of international politics. At the other end are smaller states such as Denmark and Belgium which have long had to live with much of their fate lying in the hands of others and with few opportunities to act significantly on the global stage.

However, even with variations among EU members, a CFSP should work if EU states can reach a pragmatic division of labour. In the future, if a country is reluctant to act militarily in North Africa, for example, it might make generous contributions to peacekeeping in disputes in the former Soviet empire. If a country provides a somewhat low share of the military capability of the EU, it might make up for it with a large aid programme. Clearly, too much 'free-riding' will make a sustained CFSP impossible.

A prudent approach to the CFSP suggests that often EU governments will be able to agree on the desired specific outcome of a situation, but that consensus on means to achieve that end will be more difficult, especially as more risky and costly actions came under consideration. Thus provision needs be made for even a few states to act on behalf of EU policy objectives, provided that those objectives are shared by the EU membership as a whole.

Nothing in the CFSP machinery precludes the possibility of national or multi-national initiatives by EU members, although governments seeking to lead the EU in a new direction might be prudent first to sound out privately the likely reactions of at least a sample of other members. Militarily EU governments clearly envisage WEU (and NATO) actions by *ad hoc* coalitions of the 'able and willing', as the preparations for Combined Joint Task Forces (CJTFs) demonstrate. Military actions are often the most difficult for governments to take on, given the risks and costs involved, and actions to which all WEU members will be willing to contribute must be expected to be rare. Total participation might also be militarily counter-productive given the likely problems of coordinating ten national forces (and perhaps more in future). In many aspects of foreign economic policy, however, the scope for action by some partial EU membership is smaller, given the responsibilities and powers of the Commission and the demands of the single market. Governments obviously retain control over their national foreign aid budgets.

Maastricht offers a cautious approach to the development of a CFSP, laying out broad parameters which leave scope for national interpretations and application.[9] British Eurosceptics nevertheless claim that the legal commitments of Maastricht were an error, claiming that Britain should see its main partner as the US, not Continental Europe and that Britain should concentrate on its national policy, not on developing common positions even with friendly states.

As a liberal democratic state committed to market economics, the US should have few differences of first and second level interest with its European allies. Nevertheless, the US, despite its involvement in the UN and NATO, is keen to maximise its freedom of action. Anything which narrows the range of policy choices available to the US government is even viewed as a threat to US national security.[10] The states of the EU have expressed a readiness to build a CFSP, signalling a readiness to make compromises in national stances in the interests of solidarity. There is no way, however, that the US would commit itself to the target of a common foreign and security policy with its European allies or with Britain. Historically, there have been several cases when the US has reached very different conclusions about the nature of desired action to those predominant in the UK. Former Yugoslavia provides many illustrations of this phenomenon, but the invasion of Grenada should not be overlooked, nor should the quite closely run US debate which occurred about whether to support the UK over the Falklands.

Why then should Britain not pursue its own, explicitly national foreign policy route, like the US? There are three primary reasons. First, the EU limits the UK's freedom to use economic instruments. For instance it cannot grant special access to the UK market, nor can it impose national trade restrictions except in the area of armaments (under Article 223 of the Treaty of Rome). A second consideration is that, if Britain opts for an overtly national stance, it will encourage others in Europe, including Germany and France, to act similarly, thus weakening the EU as a security community. Third, Britain does not have the resources to make a significant impact on the big issues which affect its future—the promotion of Russia as a democratic market economy, the continued commitment of the US to European security, the stabilisation of eastern Europe, the prevention of the proliferation of weapons of mass destruction, the securing of oil supplies and so on. Britain, as part of Europe, faces some alarming problems, as the Group of Experts 1994 Report for Commissioner Hans van den Broek clearly articulated.[11] With the sixth largest aid budget in the world, a professional army of 120,000, and a wealth of diplomatic expertise and contact, Britain is not a trivial power and can make a major contribution to wider efforts. Individually, however, even the major European states are likely simply to cancel each other out if they act in opposition. If the EU states say the same thing to the US, or to Russia, the message will carry weight. If a range of different messages is sent, none is likely to be clearly heard. Some would argue that British foreign policy should neglect such large issues and concentrate on promoting the wealth of the UK by helping British companies to win contracts.[12] However, a real difficulty with this approach (apart from the problem of identifying British companies in the contemporary period) is that it gambles that others will support the order necessary for economic activity to flourish. The pursuit of profit and the neglect of politics can be very expensive, as France has discovered through its relationship with Iraq. Iraq still owes France billions of dollars, mainly from arms supplied during the 1980s, and there is little immediate prospect that repayment will be forthcoming. In wider terms, it is likely that economic confidence and progress in the currently dynamic Asia–Pacific region would plummet should the US decide to abandon its security oversight of the region.

There is one other notable reason for an overtly national foreign policy—that such a policy would allow a British government to concentrate its foreign policy on postures which would be popular domestically, but which do not necessarily have any significant impact overseas. Sweden's traditional line of not supplying arms to areas of tension fits such a pattern. Clearly no government can neglect the need for domestic support for its foreign policy, but for major players in world affairs, foreign policy must be primarily about influencing the behaviour of actors outside the state, not winning votes within it. Governments must then inspire domestic endorsement of their foreign policies by persuading their publics of the benefits which will follow.

Thus consideration of the terms of Maastricht and of the nature of British national and shared interests leads to the conclusion that Britain should work with its European partners to develop common foreign policy.

Foreign policy, security and defence

In the 1980s some states, not least Ireland, preferred to exclude 'security' topics from EPC, but in 1986 the Single European Act legitimised what had become European practice: it specified that EPC could cover the economic and political aspects of security. Events in the former Warsaw Treaty area and the Gulf then presented 'foreign policy' questions with clear security and even defence dimensions. Moreover, the responses of the EC in terms of providing foreign aid, granting limited market access and contemplation of offering membership to states of the former Soviet empire obviously had important security and defence implications. Iraq's invasion of Kuwait raised a significant foreign policy issue—should the act be considered one of serious aggression? Given the agreement in the EC that it was, governments had to decide politically, economically and militarily how best to respond. If only because of the existence of the Single Market, the EC states had to give a single answer to what sorts of sanctions should be imposed. Once agreement was reached on sanctions, the immediate military question arose of whether the sanctions needed to be enforced by a blockade. In the longer term, there was the further question of when, if ever, economic sanctions would have to be supplemented by military action to liberate Kuwait. The EPC process could scarcely avoid these issues.

Not surprisingly, then, the Maastricht Treaty recognised, not just that European foreign policy cooperation could not be insulated from security topics, but also that some security issues led to the involvement of defence ministries and defence forces. At Maastricht it was settled that when foreign policy cooperation had a defence aspect, that aspect would be handled by the Western European Union.

Britain should continue to support the Maastricht commitment to develop a CFSP, with a defence dimension, because that should involve only limited qualifications of Britain's judgements of its interests and a working CFSP should be an influential force in the world.

Decision-making and the CFSP

Currently 'the EU Presidency represents the Union in CFSP matters',[13] backed by a limited staff. The Commission, with a small foreign policy directorate, DG.1A, has a voice in CFSP deliberations. Much of the political debate about the CFSP since Maastricht has concerned whether or not the CFSP process should remain inter-governmental or become integrated into Community business with the Commission taking a leading role and being accountable to the European Parliament. The main argument for such a move is that trying to secure unanimous agreement among 15 countries in time of crisis is very difficult and that a foreign policy needs some kind of central figure, backed by a significant staff, to drive it forward. The French Minister for European Affairs, Alain Lamoussure, argued at the end of February 1995 that the CFSP needed institutional innovation: 'what we need is a Lord Owen figure backed by a powerful secretariat responsible to the Council of Ministers'.[14] The Group of Experts 1994 report suggested something similar, the appointment of a 'prominent personality' supported by an information-collecting and analytical staff, which could warn of dangers and suggest action. This person would have a multi-year appointment and not be burdened by the six-month rotation which weakens the Presidency.[15]

Both the British Labour Party and the Tory government insist that CFSP must remain inter-governmental. The sound reason for keeping CFSP inter-governmental at this stage is that the Commission has done its best work in moving the Community forward only when it has been given a brief to steer the Community to specified and agreed targets. The first two such targets, reached in the first decade of the EC, were the abolition of tariffs and quotas so as to create a Common Market, and the establishment of a Common Agricultural Policy providing common prices to farmers. After arranging the entry of new members, a third target was the creation of a Single Market after 1986. In the case of foreign policy, however, the states of the EU have not specified the type of policy they want to follow or attain. The main guidance which the Commission could follow would be the unexceptionable language of the Maastricht Treaty, which provides few signals to steer CFSP through the problems of security policy in particular. Significantly the 1994 WEU document agreed in Noordwijk on a common defence policy stated repeatedly that no broad guidance for EU defence policy yet had been provided, almost hinting that EU foreign ministries should generate it.

> A full development of a common defence policy will require a common assessment and definition of the requirements and

> substance of a European defence which would first require a clear definition of the security challenges facing the European Union and a determination of appropriate responses. This will in turn depend on a judgement of the role the European Union wishes to play in the world and the contribution it wishes to make to security in its immediate neighbourhood and the wider world.[16]

In the early 1990s the governments of the Union dealt with CFSP issues on a topic-by-topic basis, covering those questions where agreement among members seemed needed or possible, or both. But the development of any kind of foreign policy philosophy, explaining how the EU views the world and its problems, and what kind of role the EU would play, had a low priority. The EC's image of itself during the Cold War was of a 'civilian power', but after 1990 the civilian power image was clearly outdated.[17] According to WEU defence ministers, a CFSP working group on security made some progress in

> developing a theoretical framework for a European security policy. It has stated general principles which could provide a basis for the further development of a CFSP. It has formulated several criteria that might apply to thinking and decision-making on possible joint actions and serve as guidance....It has made an analysis of European security interests in the new strategic context and of the risks for European security, which provides a good basis for the development of a CFSP.[18]

However, the findings of the CFSP working group have not been made public and the prime responsibility for thinking about Europe's security interests and risks may well have been passed to the large WEU group of 27 states comprising the members, associate members, observers, and associate partners in Eastern Europe.

Formulating foreign policy on a subject-by-subject basis naturally requires time to generate common views since every issue has to be treated on its own merits. In 1995 two CFSP decisions—to delay conclusion of a trade agreement with Russia because of its conduct in Chechenya and to support Japan's permanent membership of the UN Security Council,[19] demonstrated a capacity for long-term vision. But the current approach makes the EU states ill-equipped to deal with crises, when there are pressures in unexpected circumstances for quick decisions with potentially important consequences. Even single states find crisis decision-making difficult:

Japan, with its tradition of seeking consensus across almost all the political spectrum before reaching a decision, is almost in a worse position than the EU countries. But unless the EU states can develop a clearer concept of opportunities and dangers in the world, and of the EU's roles in regional and global order, the present approach to CFSP will have real limitations. Also, until there is a clear European foreign policy philosophy and broad strategy, appointing some quasi-supranational body to develop and implement CFSP would cause friction between EU members.

Policy prescriptions

Many European ministers are reluctant to become entangled in debates as to what sort of world player the EU states should be, and definition of a role will require persuasive leadership. A Europhile Britain could use its pragmatic approach to world affairs to guide a CFSP in the following directions:

1. *Give priority to the troubled regions to its east and south*:
The EU, or rather western Europe, is currently an extensive island of prosperity, peace and international cooperation surrounded by fragile democracies in central Europe and further to the east and south by poverty, political instability and actual or potential civil violence. Europe's foreign and security policy must give priority to these areas. To the east of Europe, Russia is by far the most important state to influence, given its potential both to cause problems and to strengthen cooperative approaches to the resolution of problems. (See Chapter 11)

2. *Promote economic and social development in neighbouring states in order to diminish the problems generated in these regions*:
Without a decent standard of living and reasonable prospects for economic growth, the EU's neighbours will remain incapable of sustained cooperative relations, and civil and/or international violence will remain a possibility. The main long-term thrust of western Europe's policies should be to work for economic growth, social development and democratisation in its neighbouring areas. Economic instruments of policy, including the granting of access to Europe's markets, must have a central role. British policy should be modelled on Germany and proactive '*Ostpolitik*' rather than France's cold-shoulder approach to the central and east Europeans. France,

however, is taking the lead on improving EU outreach to the EU's Mediterranean neighbours.[20]

3. *Play a role on the world stage as well as in regional contexts, not least because the US must feel that the burden of sustaining order is being equitably shared*:
On the global scale, western Europe's prosperity depends on free and secure economic activity in every continent. Also pressing for a global dimension in EU thinking is that ideas and precedents established in one region can quickly spread to others: if a government successfully commits an aggressive act in one part of the planet, governments elsewhere may well be encouraged to believe they could get away with a similar act. The concept of the indivisibility of peace has some meaning. Should one state anyway succeed to acquiring weapons of mass destruction, all others with similar ambitions will be encouraged.

The US commitment to NATO greatly facilitates provision of defence and security for western Europe, but the US will not appreciate a parochial transatlantic partner concerned only with its own immediate neighbourhood when the US is pressed to act on a wider stage. During the Cold War, in order to sustain congressional support for NATO, the US continually pressed Europeans to share the alliance burden by making a reasonable contribution to their own defence. After the Cold War, when the direct defence needs of western Europe are massively reduced, successful management of the NATO burden-sharing issue will require American confidence that western Europeans are contributing appropriately to the maintenance of order and security on a global scale.

Thus, because of the widespread economic interests of western Europe, the global nature of some issues such as proliferation and international aggression, and the nature of burden-sharing debates with the US, the CFSP must address the international system as a whole.

4. *Sustain the US commitment to European security and US engagement with issues of global order*:
In dealing with Russia, restricting the proliferation of weapons of mass destruction, preparing appropriate defences to offset proliferation developments, promoting economic and social development to Europe's east and south, and ensuring the secure flow of oil from the Gulf, the EU states can work most effectively in cooperation with the US. Insofar as the CFSP is seen as a means of bringing influence to European capitals, Washington is the capital

which Europeans will most regularly need to influence. The important message for Britain is that influence can be brought to bear on the US via the EU, especially in close cooperation with Germany, not unilaterally.

5. *Make reasonable preparations so that flexible and able military forces are available to deal with varied and difficult-to-predict contingencies. States should agree to make available between 2.5 and 3 per cent of their GDPs for defence purposes*:
The use of the military instrument in a CFSP must be viewed as exceptional, with economic and diplomatic instruments normally taking the lead. Clearly western European states cannot predict the precise demands on them to provide military forces for the wide range of operations which today and in future may support foreign and security policy. However, the EU states could develop a significant and satisfactory defence capability by making appropriate but affordable budgetary provision: the minimum that EU states should spend on defence (defined in a standard fashion) should be 2.5 per cent of GDP and the maximum, which France and Britain reached in 1994, would be 3.3 per cent of GDP. Had such an arrangement been in place in 1994, WEU states' defence spending would have been some US$23 billion higher. The varied and demanding tasks likely to be undertaken by European forces mean that professional rather than conscript troops should dominate European military structures (see Chapter 8).

If the Union is eventually going to establish an integrated, multinational intervention force, linking all WEU members,[21] it must begin soon to procure and operate some collective WEU assets. These would not involve large numbers of operating personnel, but could be expensive to acquire. There is scope for this in areas such as transportation and intelligence/reconnaissance, particularly from space. The WEU has already established a satellite data interpretation centre in Spain.

6. *Develop an effective early warning capability:*
Effective conflict prevention, resolution and management need EU governments to be well informed on developments in inter-state and intra-state relations. Conflict prevention and management initiatives should be at the centre of CFSP objectives (see Chapter 12). Given Europe's interest in peaceful, cooperative relations within and among states, the CFSP should place a high priority on conflict prevention when possible. This means that the collection of information on the

possibility of disputes turning violent, on the opportunities for the peaceful resolution of disputes, and on the means to contain conflicts geographically and in the scale of violence, should be viewed as an important task in which there are opportunities for greater cooperation among EU diplomatic and intelligence services. Providing the European Commission, or some quasi-Commission body,[22] with greater information collection and analysis roles, cannot substitute for member states' reluctance to share sensitive information they themselves have collected.

7. *Develop a clear ethical dimension*:
In order to win popular support, a CFSP therefore needs an ethical dimension: the publics of the Union need to be confident that their Union is a force for good in the world. As the *European Security Towards 2000* report suggests, EU states cannot formulate a coherent CFSP until they first agree on a set of principles they are prepared to promote and even fight for.

8. *Do not stand on principle at the expense of effectiveness*:
An ethical foreign policy does not mean that matters of priority and effectiveness will be set aside in favour of principle. The EU has finite economic, military and diplomatic resources which cannot be spread too thinly if they are to be effective. The EU cannot act everywhere, but must choose to act where the dangers are greatest and the opportunities for effective action are clear. Securing the right balance in such matters will always require judgement. However, to contribute to the deterrence of aggressive or inhuman behaviour, it is not necessary for the EU to act in all instances. Acting selectively will still leave many potential transgressors unsure of being able to get away with improper actions. Moreover, to contribute to human welfare, it is not necessary to improve every human problem: bringing economic prosperity to one society is a real advance, even if other societies have to be neglected.

Conclusion

In the Maastricht Treaty Britain legally committed itself to the development of a CFSP. This chapter has argued that a CFSP is the best vehicle for Britain's long-term influence and interests in the world, even though EU consensus on common action, especially if it has a military dimension, will often be difficult.

Britain must work diligently for a successful and positive CFSP. The lack of any clear sense of what broad role the EU states want their grouping to play in the world was the central structural weakness of the CFSP in the early 1990s. A future British government should temper its instincts for pragmatism and address matters of principle. Principles after all have played a big role in British foreign policy in the past, especially the idea that no one state should dominate the European continent. Since 1945 Britain has been obsessed by the need for a US military commitment to European security. While the value of the US commitment to Europe should not be downgraded, a more important principle for Britain in the future might be that the states of the EU should avoid the pursuit of divergent foreign policies.

7 SHARING THE BURDEN OF EUROPEAN DEFENCE

Malcolm Chalmers

The fifteen members of the European Union (EU) maintain more men and women under arms (2,090,800) than either the United States (1,650,000) or Russia (1,714,000).[1] Yet, according to one estimate, the EU states 'would struggle to deploy even three brigades to Bosnia'[2] and a 1993 study for the US Congress found that 'most are unable to contribute more than one or two battalions and associated support units to global security missions at any one time.'[3]

Europe's lack of usable military capabilities is not a result of inadequate defence budgets, but of the way in which those budgets are spent. In 1993, the 15 current members of the EU spent $177.3 billion on defence: less than the US's $297.3 billion, but more than double Russia's estimated $76.6 billion and around four times Japan's $39.7 billion.[4] Yet a high proportion of European defence spending is taken up by large conscript armies, usable only in defence of national territory. Moreover, the organisation of European armed forces on a national basis means that considerable resources are wasted on duplication. Each significant military power in Europe maintains national capabilities for research, development and production, for training and operations, for intelligence-gathering and communications. The only EU members that do not maintain the full complement of army, navy and air force are Austria and Luxembourg (neither of whom, for understandable reasons, have a navy).

The end of the Cold War exacerbated the problem of waste and duplication. When the main function of European conventional forces was to counter the threat of a large-scale attack by the Soviet Union, an evident requirement was to mobilise millions of conscript soldiers to fight on, or near, their home territory. The countries of western Europe could thus live with a proliferation of small national forces in the knowledge that in any case the United States acted as the ultimate guarantor of their security.

The United States can still be relied upon to commit itself to western Europe's defence in the event of a massive external attack, and NATO will continue to provide an essential means of expressing this guarantee. In many of the circumstances in which the use of military force is now more likely to be considered, however, US and European approaches could well differ. During the Cold War, despite their alliance against the Soviet Union, the US and its European allies often adopted divergent approaches to crises outside Europe, as can be seen in US opposition to Anglo-French intervention in Suez, European failure to support US policy in Indo-China and Central America, and a range of other issues. It is not necessary to believe that the 'glue' of the Atlantic alliance is loosening to believe that the potential remains for differences in future. Nor can European policy-makers ignore the recent increase in support within the US for a more unilateralist security policy, as evidenced most clearly in the cavalier manner in which the US treats its obligations towards the United Nations. Even the British government, in response to more than two years of snub and indifference from Washington, may be starting to listen to arguments that it 'would do better to seek special relationships elsewhere'.[5]

Experience in areas where the EU does have a common policy—such as trade—suggests that it can pursue its interests effectively when united. The contrast with security policy is stark. In the Gulf War and even closer to home in Bosnia, the countries of the EU appeared marginalised and ineffective as a result of their inability to agree a common position. If Europe is to have the capability of responding adequately to developments in the 'rim of crisis' around it—from North Africa through the Middle East to the Balkans—this will have to change.

The process of building a common policy started with the establishment of the Common Foreign and Security Policy (CFSP) in November 1993. The level of intergovernmental activity has been stepped up, and substantial progress has been made in tackling the problems surrounding NATO/WEU/EU relations. Yet early 'joint actions' have, according to one recent study, been in most cases 'poorly planned, hard to implement and disappointing both in scope and in terms of their meagre results'.[6] As a result there is growing support for efforts to improve the effectiveness of CFSP in years to come. Proposals include: more cohesive international representation for the EU (replacing the discredited 'troika' system); upgrading the EU's central analysis and evaluation capabilities; and qualified majority voting on some non-military issues. More ambitious are suggestions to adopt a common EU arms export policy and to establish a single EU seat on the UN Security Council.[7]

In 1994–95 the British government began to rethink its attitude to European defence cooperation. In September 1994, a 'twinning' agreement was signed between the British army and the French *Force D'Action Rapide*.[8] In November 1994, establishment of a joint Franco-British European Air Group was announced, to be based at High Wycombe and with a remit to help run joint peacekeeping and humanitarian missions such as those in or over Bosnia, Iraq and Rwanda.[9] In March 1995 the British government called for the WEU to be strengthened, providing it with a larger planning staff, a crisis management centre, and teams for advising ministers and handling intelligence.[10] Junior defence minister Roger Freeman favoured a 'European project office' to act as a clearing house for common procurement projects, perhaps laying the basis for a joint procurement agency in the long run. There was also support within government for scrapping Article 223 of the Treaty of Rome, which exempts defence firms from the EU's competition policies, not least as a means of benefiting Britain's relatively competitive defence manufacturers.[11]

Other measures to integrate west European forces are in place or in the pipeline. As part of the UK-led Allied Command Europe Rapid Reaction Corps (ARRC), the UK will participate in a joint airmobile division with Belgium, Holland and Germany as well as a host of other joint arrangements with European NATO members. Belgium and the Netherlands have agreed to integrate their surface fleets, with a single command structure, training and in due course procurement.[12] A WEU satellite centre at Torrejón in Spain is seen by many as the first step towards a complete family of European military satellites.[13]

Some of these measures are likely to provoke resistance from British politicians anxious to oppose any further infringement of national sovereignty. Given the limited resources available for defence in all European countries, however, these concerns will have to be balanced by the realisation that the alternative to joint action is not glorious isolation, but a growing inability to take any form of effective military action without US participation. Eurosceptics at home continue to make the government nervous about how its policies are reported, and some setbacks can be expected, but the trend is clear and seems likely to continue.

National interest and common defence

In pursuit of the objective of greater European defence cooperation, however, Britain needs to take full account of the long term economic consequences of its actions. In particular, it should resist suggestions that tie Britain into making a disproportionate commitment to collective defence. Britain's experience in the early post-war period, when the heavy burden of overseas defence commitments handicapped the domestic economy just as Germany and Japan were re-entering world markets, is a reminder of the importance of getting this balance right. In a period that proved to be critical in determining the shape of the institutions that were to dominate western Europe's 'architecture' for forty years, the British government made two decisions that were to prove extremely costly.

First, in the aftermath of the collapse of the European Defence Community in 1954, and as part of an agreement to extend the Brussels Treaty, Britain committed itself to keeping four army divisions and a tactical air force on the continent, thus reversing a long-held opposition to permanent basing in Europe. Given the nervousness that remained (especially in France) over German rearmament, and with much of the rest of western Europe still in the midst of post-war reconstruction, such a commitment was seen as necessary in order to avoid a major threat to NATO's cohesion. Yet it was to cost Britain dear, with the continental commitment taking up an increasing proportion of the defence budget as imperial commitments wound down. As late as 1990, long after the initial justifications—European poverty and French insecurity—had faded away, Britain was still supporting 70,000 army and air force personnel in Germany at an annual cost to the balance of payments of £1.4 billion.[14]

Second, largely because of its continuing attachment to a role as a world power, the British government decided in 1955 not to be a founding member of the European Economic Community. But the humiliation of Suez in 1956 and the precipitate dissolution of Empire in the years that followed dashed Britain's hopes of sustaining a global role. In 1961 Britain submitted its first application for EEC membership. It was too late, however, and a series of French vetoes kept Britain out of the Community until 1973. By this time, the Community was well established and the UK was left in the position of having to accept compromises which it had been unable to influence.

In particular, late membership meant that Britain had to accept a Community budget dominated by a Common Agricultural Policy (CAP). Some form of CAP was probably an inevitable part of the EEC, given the domestic politics of other European states. If Britain had been involved in the Community from the start, however, it might have balanced concessions in this area with measures more attuned to British interests elsewhere. As it was, Britain was obliged to become a hefty net contributor to a policy which it had played no part in devising.

Whether relatively informal (as in NATO) or formal (as in the EU), the cost-sharing arrangements for common European defence that are established in the next few years could prove as important as those agreed in the 1950s. Even if average EU defence spending fell to the German level of 2 per cent of GNP, this would still be substantially more than the agreed ceiling of 1.26 per cent of GNP on the entire existing EU budget, and more than three times the level of spending on the CAP.[15]

Were Britain again to accept a disproportionate economic burden as the price of making a common European defence arrangement work, it could prove to be as expensive (in relative terms) as the Brussels Treaty commitment made in the 1950s. Moreover, at a time when UK GNP per capita is below the EU average and still falling in relative terms, it could end hopes for a further British 'peace dividend', yet allow the UK's partners (and economic competitors) to continue to divert their resources to other, more economically productive, purposes.

The case for Britain taking a tough position on this issue is greatly strengthened because the UK is more secure from attack today than it has ever been in its history. For centuries, British governments have always sought to prevent any single hostile power from dominating the adjoining continent, in the knowledge that such a situation would pose an immediate threat to the nation's survival. Today, with the demise of the Soviet Union, that threat is no more and (given the creation of a west European 'security community') seems unlikely to return. In contrast to the period of one Cold War, the UK is now situated in one of the more tranquil corners of Europe and has comparatively less at stake in possible new crises than, say, France in Algeria or Germany in eastern Europe. To the extent that there are security implications arising from growing instability to Europe's east or south, the need to take action in response should be felt most strongly by those EU members closest to those regions.

If the countries most immediately affected do not feel that developing crises in adjoining regions justify military investment, it is far from clear why Britain should do so. The time has passed when the UK can afford to continue to give a higher priority to defence when others either do not share its threat perceptions or are prepared to 'free ride' on its efforts. By the same token, however, there may be areas—such as acceptance of asylum seekers—in which Brtian is failing to make a reasonable contribution to meeting common European responsibilities.

Britain's present contribution to European security

In response to the end of the Cold War, Britain has cut back defence spending significantly in recent years. As a proportion of national income, the defence budget fell from 3.9 per cent in 1990–91 to 3.3 per cent in 1994–5. It is due to fall further to 2.9 per cent by 1996–97 as the full savings from the 1990 'Options for Change' review, and more recently the *Front Line First* exercise, work their way through the system.[16]

Yet UK reductions have at least been matched by trends elsewhere. Between 1990–91 and 1996–97, the UK defence budget is due to fall by 17.8 per cent. But the US budget is due to fall by 25.1 per cent and Germany's budget by 25.3 per cent. France, of NATO's political powers, was the last to inaugurate reductions in defence spending.

As a result the 2.9 per cent of British GDP projected to be taken by defence in 1996–97 'will be higher...than virtually all other European NATO countries are today and they are going down'.[17] In proportional terms, Britain remains the biggest spender in the EU with the exception of Greece, with the latter's budget dominated by the requirements of pursuing its long rivalry with Turkey. Italy—situated next to both the Balkans and North Africa, and with a total GNP now larger than the UK's—has a defence budget only 60 per cent of the British level. Spain, despite an economy half the size of the UK's, spends only a fifth of the amount on defence.[18]

Moreover the raw budgetary figures tell only part of the story. For the UK also makes a particularly large contribution to collective defence in those areas which are likely to be of greatest value in the post-Cold War environment. Despite ranking only fourth in economic strength within the EU, Britain now probably ranks just ahead of, or equal to, France as the EU's most capable all-round military power.

The UK has had to devote more resources to internal security than most other EU states because of the situation in Northern Ireland, and it is one of only two nuclear weapon states in the Union. Yet it is also able to field a range of forces for conventional warfare that compares favourably with any of its European allies. The army holds the command of NATO's Rapid Reaction Corps and contributes more forces to it than any other nation. Although it cannot match France's nuclear-powered aircraft carrier, the Royal Navy has a larger frigate and destroyer fleet, and probably a more capable submarine fleet, and is clearly superior to any of the other EU navies. Finally, as a result of the ambitious procurement programmes of the 1980s and early 1990s, the RAF's combat aircraft fleet now consists predominantly of recently introduced Harrier GR7, Sentry and Tornado models: making it one of the most modern, and most capable, air forces in western Europe.

In addition, the UK continues to do more to modernise these forces and make them capable of fulfilling new roles. Despite the fact that its procurement budget has fallen by 19 per cent in real terms between 1990–91 and 1994–95, the UK continues to spend 49 per cent more on equipment than Germany.[19] As a result of its relatively large procurement budget, the UK often takes the biggest share in collaborative projects, for example in the UK–France–Italy Horizon frigate, the UK–Italy EH-101 helicopter, and most notably in the UK–Germany–Italy–Spain Eurofighter programme.

Even in land systems, the army's continuing modernisation programme means that, despite the reduction in total army personnel in Germany from 56,000 to 23,000, its combat power will be reduced by only 20 per cent.[20] At a time when many countries, including the US and Germany, have no further plans for tank purchases, the UK is going ahead with plans to replace completely its existing Chieftain and Challenger 1 tanks with the new improved Challenger 2 model.

This impressive record is in part a result of the success of the efficiency reforms introduced into the Ministry of Defence since the mid-1980s. But it may also reflect the fact that the UK has not had to tie up a large proportion of its defence resources in maintaining large conscript forces. Such forces may still fulfil useful social and even political roles. But conscription is less and less appropriate in military terms, and both Belgium and the Netherlands have now announced their intention to move towards much smaller, but all-volunteer, forces.

Nor is it even clear that conscription saves money, despite the fact that it is sometimes argued that Britain spends more on defence because it has to pay personnel who are employed for free (or very little) in conscript forces.[21] One difficulty with this argument is that the UK spends a lower proportion of its defence budget on personnel than any of its European allies:[22] the opposite of what one would expect if conscript armed forces were simply lower-cost versions of their volunteer equivalents.

Finally, it is possible in theory that other countries may be compensating for Britain's disproportionate defence effort by spending more on other 'international public goods' such as overseas aid. Here Britain's record has been poor, especially in the period since 1979. The 0.28 per cent of GNP allocated to aid in 1994–95 falls far short of the UN's target of 0.7 per cent, which the government has repeatedly promised to reach. It also compares unfavourably with some of the smaller EU members such as Denmark (1.0 per cent), the Netherlands (0.79 per cent), and Sweden (0.9 per cent). Yet, compared with other large and medium-sized powers, Britain's record in recent years is by no means the worst. The UK continues to spend more than the US (0.2 per cent) and Italy (0.19 per cent) and about the same as Japan (0.3 per cent), Germany (0.33 per cent) and Spain (0.26 per cent). Of the medium powers, only France (0.6 per cent) spends significantly more than the UK. Given the changing nature of Europe's security concerns, there is a strong case to be made for more resources to be shifted into development aid. Whether this takes place or not, however, it is clear that Britain's low aid budget cannot be used as a means of justifying a higher-than-average defence budget.[23]

Towards a European burden-sharing agreement

It is in the UK's interests to encourage development of a strong common European defence policy. But the flipside of responsibility-sharing—which will inevitably bring with it some dilution of UK and French influence—is greater burden-sharing. A 'no representation without contribution' policy would not be immediately welcomed by some of Britain's partners. But it would inject realism into the discussion of a common policy, and would help other members understand that they too will have to make some difficult adjustments to realise such a policy.

A common policy would also enable the UK to press for better access for its defence industries to the protected markets of many of its allies. One of the reasons why the EU spends so much on defence, but produces relatively little for it, is the proliferation of 'penny packet' forces and arms industries. The UK should be at the forefront of the argument for a more integrated policy as a means of getting value for money in European defence.

In moving towards more equitable funding of common defence, one possible starting point might be for the WEU to agree to fund operations carried out on its behalf by its members. A degree of cross-financing already takes place on an informal basis, for example the German contribution of £275 million to the UK during the Gulf War. But this could be formalised so that, in addition to participating in a decision for an operation, all WEU members shared in the costs of that operation.

Yet this can only begin to tackle the problem. For, except in circumstances of warfare sustained over a period of years, the additional costs of operations are dwarfed by the ongoing costs of maintaining and equipping the forces that are used in these operations. Thus, the total additional cost of the Gulf War to Britain was a one-off outlay of around £2.5 billion,[24] but the annual defence budget for 1990–91 was £22.3 billion. From a financial point of view, therefore, the arrangements for sharing the peacetime costs of forces are likely to be much more significant than those for operations.

A recent study into the future development of the CFSP suggested one possible way forward. After suggesting that the Union Treaty 'should explicitly provide for the build-up of Eurocorps and other multinational units designated for the WEU into a sizeable European intervention force', it argued that:

> It should further be made clear in the Treaty that the intervention force, by definition at the service of the EU...must from the outset receive political and financial backing from those Member States which do not wish to participate.[25]

In line with this recommendation, multinational forces designated as being at the service of the WEU, including Eurocorps, the Anglo-Dutch amphibious force, Multinational Division (Central), and possibly other European elements of NATO's Rapid Reaction Corps could be financed in the same manner as other elements of the EU budget, with contributions calculated on the basis of GNP and/or VAT revenues.

Such an approach could bring substantial benefits to the UK taxpayer, who could expect to save several billion pounds a year if it were to be applied to the majority of its forces. Moreover, it would not raise insoluble issues of sovereignty and control, any more than do Japanese payments to the US for its forces in Japan, or German offset payments to the UK did during the 1960s. The main obstacle is likely to be the resistance of Britain's European partners. At a time when several of the EU's lowest military spenders (such as Belgium, Italy and Spain) are cutting their own defence budgets further, in part in response to the need to curb budget deficits in preparation for a common European currency, there is unlikely to be much political mileage in supporting substantial transfers to the armed forces of fellow EU members.

Even if other EU states are undervaluing the potential importance of military force, it is not clear that it is in the British national interest to make a disproportionate contribution to common defence if other members of the Union do not value that contribution enough to support it financially. Other things being equal, one would expect a country like Britain with below-EU average income situated in one of the safest parts of Europe to spend less of its national income on defence compared with the rest of the EU, not more.

Yet the need for an increased focus on intra-European burden-sharing as an issue is not simply a matter of saving the British taxpayer money. It is also a necessary component of a genuinely integrated common defence policy. As long as Britain and France are prepared to take the lead by spending more on defence, others will be content to let them get on with it, just as they relied on the US to provide a security 'umbrella' for them during the Cold War. Even if they are involved in an elaborate architecture of consultation, they will not take real responsibility for common policies and they will not take on the real costs of common commitments. Some may even see the disproportionate costs of defence for Britain and France as the manifestation of continuing post-imperial delusions.

Even if neither increased influence in Europe nor the 'special relationship' with the US justifies Britain's defence budget, it is sometimes argued that Britain needs to spend more on defence because of its permanent membership of the UN Security Council.

It is far from clear that the additional national pride engendered by possession of this seat is sufficient to justify the additional one per cent of GNP that the UK spends on defence over and above the European average. In any case, the argument is based on a false premise. While the value of a permanent seat may well be diluted in future by extending permanent membership to Germany and Japan (incidentally, further eroding the link between military contribution and political representation), there is no realistic possibility that Britain will lose its seat.

Implications for the next British defence review

For the foreseeable future, most EU defence spending will be financed nationally. If the gap between the UK's defence burden and those of its EU partners is to be reduced, therefore, it will require a thorough review of defence commitments by the British government. Such a review cannot, however, be undertaken in isolation. It will require a broader review of Foreign and Security Policy so that a defence review is based, not only on the need for financial savings or on a narrow computation of military 'requirements', but also on a sense of Britain's new security interests in an era very different from the militarised peace—or 'Cold War'—from which the world has now emerged. In particular, a defence review will need to be closely linked to the government's policy towards the European Union. The government will need to articulate clearly how it would like the CFSP to develop, and make clear that a defence review would be undertaken in close consultation with Britain's European partners. It is difficult to predict in advance the savings that might result from such a review. These will depend in part on developments in and around the former Soviet Union: still the most likely source of instability for the EU. If Russian military capability continues to decline, the UK and its allies can also afford to relax their guard further. On the other hand, if significant new threats to EU members, or to potential EU members such as Poland, were to emerge, EU force reductions might have to be put on hold. In this latter case, rather than involving further reductions in UK defence spending, a more equitable burden-sharing between Britain and its allies could be for others to increase their defence efforts while Britain held its own contribution steady.

Absent of new threats, progress towards a fairer distribution of defence burdens within the EU is likely to be achieved by budget reductions at around the pace (3–4 per cent per annum) of 1990–95.[26] The experience of this period suggests that it is possible to generate sustainable savings by a combination of targeted force reductions and a continuing programme of measures designed to increase the efficiency of the forces that remain. If, in broad terms, the government sought to make a further reduction comparable in magnitude to that of the last five years, it could expect to achieve a saving of around £4 billion per annum by the year 2002, bringing UK defence spending down to around 2 per cent of GNP by that year.

The starting point for a new defence review—even if conducted in close cooperation with EU allies—would have to be the continuing need to fulfil unavoidable national military commitments, defined as those that are uniquely, or primarily, British. No other country can be expected to share substantially in providing forces for these commitments. These will include a number of minor tasks, including fishery protection, search and rescue, and ceremonial duties. But, after the withdrawal of forces from Hong Kong in 1997, only two commitments are of any significance for force planning purposes. The first is to defend the Falkland Islands against possible Argentine attack. As long as the UK maintains a garrison in the Islands and a demonstrated capability for immediate defence and rapid reinforcement, a guaranteed capability to retake the Islands—problematic even in 1982—may be less important. Even so, the ability to reinforce a garrison at such long range will continue to require capabilities that might otherwise have been given a lower priority.

The second significant national commitment is the armed forces' role in support of the civil power in combating terrorism in Northern Ireland, together with associated counter-terrorist roles on the mainland. Until the 1994 ceasefire, 19,000 service personnel were based in Northern Ireland, and many more were needed to sustain this commitment because of the policy of troop rotation. Until further political progress is made in the discussions of the future of Northern Ireland, the need to guard against the possibility of a renewed terrorist threat will severely limit the potential for further economies in the army. On the other hand, if a lasting settlement is achieved, it should be possible to reduce the overall requirement for army personnel, even if some allowance is made for a larger commitment of infantry to UN peacekeeping.

Yet, even if there is no lasting settlement in Northern Ireland, the forces to support uniquely national commitments probably account for no more than a quarter of the current defence budget of £24 billion. And many of the forces that are included in this total—for example those needed for reinforcing the Falklands—can also be used to contribute to collective NATO or WEU missions.

Decisions as to the nature, and extent, of the savings that might be made in the forces that remain—i.e. those that do not support a uniquely national commitment—will require consideration of many factors that can only be determined with full access to all the relevant technical data, and with the opportunity to conduct extensive consultation with allies. But it might be useful to illustrate the potential for savings in a new defence review by looking at three of the key issues facing defence planners: the future of the army in Germany, the need for UK air defences, and the necessary size of the frigate and destroyer fleet.

The army in Germany

A defence review will undoubtedly want to look again at the size of the contribution made by the army to NATO's Rapid Reaction Corps.[27] In addition to providing 60 per cent of HQ staff, considerable combat support (including armoured reconnaissance), and an airmobile brigade for Multinational Division (Central), the UK also provides two divisions to the Corps: the entirely national 1st Armoured Division, based in Germany, and 3rd Division, which includes three UK-based British brigades plus one Italian armoured brigade.[28]

There is likely to be particular debate about the role of 1st Armoured Division. A defence review may question whether it still makes sense to devote such a significant level of resources to maintaining an army presence in Germany now that the military rationale for that presence—forward deployed Soviet troops—has gone. Keeping almost 30,000 troops in Germany, along with associated civil personnel and families, costs the UK substantial amounts of foreign exchange every year. Even excluding the troops deployed in Northern Ireland, the UK will have a higher proportion of its army deployed outside national territory than any other major army in the world, imposing both an additional financial burden and considerable disruption for personnel.

Given the importance of Britain's relationship with Germany, any decisions about British troops there must be taken only after discussion with the German government. But, given its common membership in both the EU and NATO, the UK's commitment to the defence of Germany—were it ever attacked—is hardly under question, and could be reinforced by ongoing preparations for reinforcement from the UK should a threat from the east re-emerge. Nor should the argument that Britain needs to keep forces in place in order to 'restrain' Germany—which may have made sense 40 years ago—be given any credence. Even if Germany were to change, the idea that a single British division on the Rhine could play any positive role in influencing German politics is absurd.

The case for reviewing the army presence in Germany is strengthened by the debatable value of the UK replicating—albeit on a smaller scale—the capabilities of the US and Germany for large-scale armoured warfare. (This is the primary role of the UK's eight armoured regiments in Germany.) The UK has little comparative advantage in this area, and it is not relevant to any of its remaining national commitments. The army can still make a significant contribution to collective military capabilities in other areas, and there is a case for retaining some armoured capability for smaller-scale operations. Given limited resources, however, it might be better to accept that it should not attempt to do everything that its larger counterparts can.

Air defence and the Eurofighter

The prospect of conventional bomber attack against the UK is now very low. As a consequence, the government is already reducing the number of squadrons and aircraft devoted to this commitment. But there may be scope for further economies in this area. Air defence capabilities will still be needed for power projection (for example to Bosnia or the Falklands). But much of the existing UK-based infrastructure would appear to be much less relevant than in the past. A thorough review is needed to determine the size of the forces needed for future air defence.

As part of that re-examination, the government should reassess the current requirement to buy 250 Eurofighters at a projected capital cost of £15 billion, and an estimated lifetime cost (including operations) of at least £22 billion.[29] On current plans, the EF2000 programme and its associated weapon systems will require a doubling of the air force equipment programme between 1994 and the end of the century.[30] It is hard to see how this level of increase can be sustained without either a reversal of recent defence spending cuts or drastic economies in the two other services.

Yet, as the need for air defences falls, it is increasingly difficult to understand why the UK needs to buy 250 aircraft while Germany is buying only 140, Italy 130 and Spain 87.[31] The UK's defence needs have changed drastically since the project was originally conceived. A defence review might consider whether these needs could not still be adequately met with a significantly reduced purchase of aircraft.

The size of the frigate and destroyer fleet

As part of the force reductions of the early 1990s, the size of the frigate and destroyer fleet has been reduced from 50 to 36. In the new strategic environment, however, this reduction may not have gone far enough. Specialist forces in which the UK has particular strengths, such as amphibious forces and ASW carriers, give the UK a range of military options that could well be of value in future crises. By comparison, the military options forgone by reducing the number of frigates and destroyers to around 25 may be relatively modest. The virtual end of a Soviet naval threat has released the Royal Navy's frigates and destroyers from what was, until recently, their primary mission: anti-submarine warfare in the North Atlantic. Making a modest contribution to global peacekeeping and peace enforcement missions is a legitimate role for British forces. But this is a role which the Royal Navy can share with other navies, both from Europe and from elsewhere. It is not a role that justifies Britain retaining a larger fleet of frigates and destroyers than any other EU member state.

A reduction in the number of frigates and destroyers would, *inter alia*, allow the government to reconsider the need to replace the ageing Type 42 destroyer on a one-for-one basis. Currently, the UK is due to purchase 12 of the proposed 'Horizon' frigates for this purpose, while collaborative partners France and Italy together intend to buy no more than eight.[32]

Table 7.1 European defence spending as a proportion of GDP

country European Union members	per cent GNP spent on defence 1993	1993 GDP per head at PPP	Active armed forces personnel 1994
Luxembourg	1.2	20,900	800
Germany	2.0	20,800	367,000
France	3.4	19,500	410,000
Denmark	2.0	19,100	27,000
Austria	0.9	18,800	51,000
Belgium	1.8	18,400	63,000
Italy	2.1	18,000	322,000
Netherlands	2.3	18,000	71,000
Sweden	2.5	17,500	64,000
UK	**3.7**	**17,300**	**254,000**
Finland	1.9	16,000	31,000
Spain	1.5	13,400	207,000
Ireland	1.1	12,600	13,000
Portugal	2.9	8,800	51,000
Greece	5.4	8,000	159,000

Source: IISS, Military Balance 1994–1995

Conclusion

In the 1950s, despite the damage to its own economy, the UK accepted a disproportionate share of the costs of Western defence in order to secure the common objective of collective defence. In the 1970s, it was willing to accept a Common Agricultural Policy from which it was bound to lose as the necessary price to be paid for the broader benefits of European Community membership. In the 1990s, as Europe moves towards the adoption of a common defence policy, there is a danger that, in order to see this new policy succeed, the British government will allow history to repeat itself.

It will still be important that the government provide whatever forces are needed to meet uniquely national military commitments. Yet Britain's national income per head is now lower than those of Germany, France and Italy, and a halt to the relative decline of the last forty years will not be possible without a single-minded concentration of resources on national economic revitalisation. In these circumstances, and given the relative security of the British Isles, it is hard to see the case for the UK allowing itself once again to provide a disproportionate share of collective defence requirements.

8 THE EUROPEAN DEFENCE INDUSTRIAL BASE[1]

Philip Gummett

Defence planning has become more difficult since the end of bipolarity. Until the late 1980s, the size and shape of western European defence activities reflected the specific geopolitical circumstances of the Cold War. As these circumstances changed, the defence sector suffered not just a loss of demand, but a loss of coherence. The military strategies and force structures pertinent to defending against massed attacks from the east lost their relevance. It became less clear how to organise military forces in response to the more diffuse set of risks that have emerged.

Countries now face the problems of identifying requirements in a coherent fashion, and then of finding the necessary funds at a time of public pressure for cuts. When 'the threat' came clearly from the Soviet Union, it was possible to plan fairly straightforwardly. Moreover, because of the scale of forces assembled to deal with that threat, lesser defence problems could be managed with some configuration of forces drawn from the large total. But the single large threat has now been replaced by multiple smaller, but less clear-cut, 'risks'. Not only is it difficult to know what to plan for, but the forces available will be much smaller than in the past, thus reducing the scope for withdrawing elements of the total to meet particular eventualities. If it could be confidently assumed by each country that it would meet all conceivable eventualities in the company of the same fixed set of allies, then a division of labour might be agreed. But even this seems implausible at present. Hence the profound uncertainty that grips defence planners and, by extension, those responsible for planning defence research and development (R&D). This chapter reviews the future of British and European military technological and industrial capacity by seeking to answer the following questions:

- How can defence equipment be procured on an affordable

basis in the light of reduced markets and technological change?

- Can the relations between defence and civil science and technology be conceived and organised so as to assist with the problem of affordable defence equipment, while at the same time using defence procurement for the greater economic good?
- Can we reposition the expertise and capacity that resides in the defence sector, so as to retain defence capacity where it is needed, while also moving away from the position of having highly defence-dependent firms who then constitute major problems for governments when the defence market is in decline?
- What are the implications of developments elsewhere in Europe for British decisions on this subject?

Implications of defence changes for R&D

All the countries that were involved in the former Cold War are in the throes of making deep cuts in their defence forces and supporting structures. The essential point to note, however, is how little difference these cuts are making to the shape of the inventory of new equipment requirements of the larger European states. We must note the important exceptions of the new emphasis on Command, Control, Communications and Intelligence (C^3I), and defences against weapons of mass destruction, offset by some reductions in nuclear weapons programmes, more so in France than in Britain. But most major programmes are being kept in place, albeit stretched out in time and often with a reduced number of units.

At the level of research and development, governments responded initially to the post-Cold War conditions by asserting the importance of maintaining high levels of technological competence, if not necessarily pushing this through into finished products. For example, in a speech in September 1991, the then British Minister of State for Defence Procurement, Alan Clark, referring to the importance of developing a 'reconstitution capability' as an element of defence strategy, suggested that 'we might also consider whether to give R&D greater priority within the defence budget.'[2] He added that he was not proposing a flood of new money, but rather was wondering

> whether we should not be researching new technologies and demonstrating them, while not automatically taking them into Full

Development as before. We would therefore be thinking in terms of giving pure research, and technology demonstration, greater priority.

This suggested that defence R&D would not necessarily be cut at the same rate as the rest of the defence budget.

The overall turbulence of defence policy is making it difficult for clear lines of R&D policy to emerge. In addition, various inadequacies of statistics, such as the very limited projections which most countries are prepared to publish, together with lack of information about industrial spending, and difficulties of accounting for dual-use projects, also complicate the task of analysis. Nevertheless, it does seem that western European governments are indeed protecting their defence research, at least relative to the rate of reduction of defence budgets in general.[3]

Britain, France and Germany

- In *Britain*, defence R&D is planned to fall between 1987 and 2000 by about one-third. However, within this total, defence *research* is expected to fall by only 15 per cent by 1998.[4] Spending on defence as a whole, in contrast, is expected to fall by 14 per cent in real terms between 1992–93 and 1996–97 alone.[5] Within these totals, there is to be a 'modest though noticeable shift away from areas of research on platforms (such as ships and aircraft) towards research on weapons and sensors'.[6] The 1994 defence costs study, however, refers to what *appear* to be *additional* planned savings of £40–50 million per year (against a 1993–94 research estimate of £630 million), and 1,300 job reductions by 2000, resulting from reorganisation of science and technology services and redefinition of research funded by MoD.[7]
- In *France*, the defence minister aimed in 1992 to maintain defence research at at least 6 per cent of what, under the 1992 Loi de Programmation, was planned to be a slowly declining equipment budget.[8] Within this envelope, there was to be a marked switch from nuclear to space and intelligence related activities, and redundancies were announced at the main French nuclear weapons research establishment, though even here the cuts appeared to be concentrated in administration and

production rather than research.[9] There is no clear evidence of significant cuts otherwise in conventional defence R&D. Overall, French spending on defence R&D has remained for several years at about 33–4 per cent of procurement expenditure.

- In *Germany*, where the absolute level of military R&D spending has of course been much lower than in the UK and France, reductions are planned, but these are likely to be very much less than planned cuts in procurement. Thus, whereas procurement spending is expected to fall by about half between 1989 and 1994 (a much larger fall than for the UK), R&D spending is expected to fall by only about one sixth.

Industrial developments

The future west European capability in defence technology depends not only on the policies of governments, but also on the decisions of companies. This is not because of industry's role as a sponsor of defence technology, where it plays a minor part compared to governments, but rather because in every country it is in industry rather than government that most of the *physical* capability is to be found.

Industry responds to different pressures from those operating on governments. Accordingly, while European governments have gradually revised their defence policies over the past several years, defence industrial companies have moved ahead more rapidly in order to try to maintain their own viability. These developments are important to the future of European foreign and security policy, at both national and EU levels:

- they influence the defence capability of Europe as a whole, as well as that of individual states;
- they affect the economic security of states, both in terms of direct effects (such as jobs) and by altering the general attractiveness of a country, or of regions of a country, for high technology investments, according to the 'cluster' theory of Porter, and the arguments of Reich[10] over the importance of maintaining a skilled workforce as a magnet for such investments;
- for both political and economic reasons, therefore, the

protection of national defence industries is a natural concern of most governments, but this concern runs against those currents in European politics that are flowing towards greater integration;

- they influence the scope for shifting the balance of European technological capabilities more in the direction of dual-use technologies, as discussed further below.

West Europe's defence companies are engaged in major restructuring, within and across borders, which is likely to accelerate still further. Skilled people are being displaced from their jobs, with little comparable work to go to. A 1990 SIPRI estimate suggested that between 350,000 and 500,000 jobs in the defence and related industries could be lost in the European NATO countries between 1990 and 1995,[11] and later SIPRI data suggested that even the higher of those figures might be an underestimate.[12] This could result in a substantial reduction in the scientific and technological labour pool available in Europe. While there would clearly be little point in maintaining capabilities for which there is no demand, the abruptness and (often) geographical concentration of defence cuts can mean that there is little chance of any alternative demand within a short time period.

It is important to note that, although accelerated by the end of the Cold War, the origins of this industrial turbulence predate the revolutions of 1988 and 1989, and the dissolution of the Soviet Union.[13] Even without Mr Gorbachev, profound changes would have occurred at the defence industrial level. These arose from a general overcapacity of production, a shifting balance of leadership between defence and civil technologies, the indirect effects of the Single European Act, and a concern within defence ministries about the competitiveness of European defence firms *vis-à-vis* their United States counterparts. The last of these concerns has grown even more acute following the Gulf War, as the higher standard of US arms production became abundantly clear, and as US arms exporters fixed their gaze more firmly on both Europe and the markets to which European firms themselves have traditionally exported.

It deserves emphasis that *international* restructuring has come to the defence sector later than to the rest of industry, with defence still today being regarded as a sector of unusual strategic significance by governments, who therefore actively scrutinise foreign interventions. International restructuring has, nevertheless, been particularly evident in the electronics and aerospace sectors. Restructuring of the warship and land equipment industries has tended so far to remain within national boundaries, although even this is changing.

Two main patterns of activity can be observed. In one of these major defence contractors take over second-tier firms in other countries. Of greater significance, however, have been the moves to establish international joint ventures between the major players on the European scene, particularly between British and French, and French and German, firms. These present the prospect of radical reshaping of the industry, which might result in the formation of a few large multinational 'clusters' of fixed composition, but more probably perhaps will take the form of fluid coalitions of opportunity between divisions of the main defence firms.[14]

These developments will affect not only the size and distribution of Europe's defence industrial capabilities, but also the balance of technological power within Europe and the competitiveness of Europe as a whole *vis-à-vis* the USA and Japan. There is, however, a danger that individual companies will take decisions according to the logic of their own positions that, in the aggregate, may turn out not to be in the best interests either of their own states or of Europe as a whole.

New civil–military technological relations: the Japanese model

The management of technology for defence purposes therefore faces several challenges. It will be less well funded than in the past. Production runs will be shorter, encouraging international collaboration and/or advanced manufacturing methods in order to keep prices within reach. The industry will be increasingly internationalised, making national control of technological assets more difficult. And the traditional post-1945 assumption that military technology was more advanced than civil is being reversed, in areas from electronics to structural materials, as the commercial sector, with its vastly greater size and its growing performance standards, is increasingly making the technological running.[15] This last point means, first, that defence suppliers have to learn how to draw more actively on the civil technology base (and may indeed be threatened by traditionally civil suppliers in the competition for defence contracts); and, second, that the process of supply is becoming increasingly internationalised, not to say globalised, not only at the level of the prime contractors, but right down to the level of suppliers of components and materials.

How, in these circumstances, can governments and firms maintain technological dynamism? How can defence equipment of adequate quality be acquired at an affordable price?

The way ahead depends on recognising the need to switch to a new defence technological 'paradigm', the prime example of which is, of course, Japan. Japan has become a major producer of high quality defence equipment (as well, of course, as a prime supplier of components to western defence equipment), without having developed a specialised defence industry; defence production without a defence industry.[16] Defence production is managed as a small part of the work of large, technologically dynamic companies, whose main competencies and markets lie elsewhere.[17]

The best account of the Japanese case is that by Samuels,[18] who shows that Japan has a long history of military technological development. This 'technonationalism', as Samuels calls it, comprised then (and still does) three elements: import-substituting *indigenisation* (the identification and acquisition of foreign technology in order to stimulate local development); *diffusion* of this knowhow throughout the country; and *nurturance* of a capacity to innovate and manufacture. Hence, fact-finding, technology-acquiring tours by visiting Japanese are nothing new. There is a long history of sending engineers abroad to acquire new techniques and apply them in Japan. There is also a long history of active government intervention in this process, not least in the military sector.

But, goes the popular myth, surely this history applies only to the period up to 1945 and not beyond? Here again Samuels destroys the myth. Japan's defence options were abruptly altered by the outcome of the Second World War, but arms production attracted considerable attention from economic planners and business leaders even in the early 1950s. In fact, says Samuels, 'military procurement was an engine of Japan's postwar reconstruction and ever since has been an important source of technology.' Crucially, however, he continues: 'Perhaps more than any other nation, Japan has successfully embedded a defense industry in a commercial economy.' Samuels argues that:

The fundamental lesson from the Japanese experience is that a full-

> spectrum commercial capability helps defense production as much as focused defense industrial policies. Each of the pieces, up and downstream, meshes together, and the diverse commercial economy that results is a huge 'knowledge generator' for society.

How different the debate over the future of defence industries in the west and in Russia would be were they starting from such a position.

Instead, Britain starts from the position of having an arms industry that is a specialised enclave within the industrial economy. That industry has achieved its current status under the old technological and industrial paradigm, and in conditions which enabled it to become firmly embedded within national political and economic structures. It is therefore relatively well placed to seek to buttress its existing position. The emerging new technological paradigm, most clearly exemplified in Japan, threatens, however, to undermine Britain's position. It does so not only in terms of the threat to the international competitiveness of the firms in question, especially with respect to any civil interests that they may have (where they enjoy less government support), but also through the fact that Britain's defence ministry, like its counterparts elsewhere, can be expected to look at new, cheaper, possibilities for meeting its needs.

Such an approach would also help to alleviate another potential problem that could arise from the present wave of industrial concentration, namely, how to maintain competition, if not within individual countries (which is increasingly implausible), then at least within Europe as a whole. By relying less on a specialised defence sector, and more on a generally strong technological and industrial base, the prospects for so doing would be improved.

Dual-use technologies

This discussion is now organised in several countries around the concept of dual-use technologies. Thus there has already been extensive analysis of this concept in the USA.[19] Since the election of President Clinton, moreover, the subject has moved from the level of analysis to the level of policy. It has become a central pillar of defence technology orthodoxy,[20] with the 1995 Department of Defense budget containing over $2 billion for dual-use work. Similar emphases are apparent also in Russia, though the follow-through into practice is rendered more difficult by the problems of limited demand for civil high-technology.[21]

What is the scope in Europe for serious development of dual-use technologies, as the basis for both defence and civil activities? To the extent that European governments and firms do address the concept of dual-use technologies, do they do so in the spirit of aiming towards the 'Japanese model', or of simply seeking new forms of financial support for a besieged defence sector? Are European firms capable organisationally of working in the 'Japanese' way, or is the habit (found more strongly in some countries than in others, it must be said) of organising defence production separately from civil too strongly entrenched? A variety of positions can be seen in Europe, deriving from different conceptions of the status of defence affairs, at one level, and of the relations between the defence and civil sectors, at another.

Britain is becoming increasingly active on this subject. Industry is pressing strongly for a greater national effort in the development of dual-use technologies, notably through the programme operating under this title, sponsored by the CBI on the initiative of British Aerospace.[22] Industry has also called for a wider national role to be played by the Defence Research Agency (DRA; now part of the Defence Evaluation and Research Agency, DERA) in the form of support for the overall science and technology base rather than a narrow concentration on defence purposes only. The same theme arises in a paper submitted to the MoD by the Defence Industries Council in late 1993[23] which calls for a study of future technological requirements not unlike that recently conducted in France for the Commissariat général du plan[24] (see below), and is emerging from the conclusions of the Defence and Aerospace Panel in the Technology Foresight Programme being run by the Office of Science and Technology.

These moves run in parallel with measures being taken in UK government in general, and the DRA in particular, to develop new Dual-Use Technology Centres, and related initiatives which will align the programmes of the DRA more closely with those of industry.[25] However, just as it is not clear how far the outcome of the industrial initiatives will in practice be confined to bolstering existing defence firms, so the motives behind the governmental initiatives are also mixed. Specifically, they arise at least as much from the DRA's need to cut its fixed costs by rationalising its facilities onto fewer sites, while also attracting cooperative industrial support for these facilities, as from any clear policy lead by the MoD. In September 1991 Defence Minister Alan Clark said:

> In the future I see the MoD making increased use of commercial off-the-shelf technology wherever we can. Our R&D will be increasingly targeted on producing military derivatives of basically civil technology, and on those areas where civil technology is not suitable or simply does not exist. In the constantly extending border area between military and civil technology there will be scope for MoD to become a more eclectic—though discriminating—patron and customer of work with potential for civil application.[26]

However, this was something of an exceptional remark. Even today, the UK MoD insists that, while it is content for there to be civil benefit from defence programmes, defence funds can only be spent on activities for which there is a clear and primary defence need. This obviously inhibits the scope for developing the dual-use approach.

Much evidence on this score was given in 1994 to the enquiry into the DRA by the House of Lords Select Committee on Science and Technology, resulting in a recommendation by the Committee that the DRA be encouraged and enabled to play a more prominent role in national wealth creation.[27] It is worth adding that the Ministry of Defence and the DRA are also being increasingly drawn into wider discussions of the scope and direction of their research programmes by virtue of their involvement in the machinery of science and technology policy in the UK. Specifically, they now have to make statements of overall strategy for the annual *Forward Look of Government-funded Science, Engineering and Technology*, and have contributed actively to the Technology Foresight exercise, which offers some possibility of harmonising MoD's requirements with other national scientific and technological efforts.[28] Nevertheless, defence–civil relations remain more distant in Britain than in some other European countries and, as again was seen in evidence to the House of Lords enquiry, DTI's capacity to offer financial rather than simply moral support to dual-use programmes at DRA is declining rapidly.

A similar debate has also been under way for some time in France, but in a tone altogether more positive than in the UK. France has the largest military industry in western Europe, mainly still under state control and located within distinct national champions, but with an increasing proportion of it facing privatisation. This industrial strength is supported by a strategy of independence which, however, gets harder by the year to maintain. France is already dependent on foreign suppliers for various materials and components, and partly in response to this position plays an extremely active role in European industrial collaboration.

French expenditure on military R&D has remained fairly stable as a percentage of procurement expenditure since 1982, varying between 31 per cent and 36 per cent, but typically being around 33–5 per cent. The Ministry of Defence also funds more research in companies than any other French ministry. The Ministry concentrates its technical expertise on armaments questions within the *Délégation Générale pour l'Armement* (DGA), which oversees the research programme, performance of which is distributed between industry (63 per cent), universities and research centres of other ministries (7 per cent), and research organisations under the supervision of the Ministry of Defence (30 per cent). The figure for intramural research (*not* development) expenditure in France of only 30 per cent compares with one of more than double that in the UK.[29] The consequent predominance of industry in the performance of defence research, together with the volume of contracts placed by the DGA directly with firms, is seen in France as creating a dense network of links between public and private laboratories. This, some say, is 'a guarantee of the best use of dual-use technologies'.[30]

More than this, however, is the explicit attention that has been given in France to the development of dual-use technologies. Thus, for several years there have existed large annual meetings to draw together defence and civil scientists (so-called *Entretiens science et défense*). Civil agencies, such as the national space agency (CNES) are much more closely involved with defence programmes than would be the case in, say, the UK.

The tone of the French outlook can be seen in the following quotation, taken from an editorial in the house journal of the French DGA, in introducing in 1991 a special issue on '*Les Technologies Duales*':

> The time is past when the constraints of secrecy or of specificity could lead to defence research being conducted in isolation, hoping perhaps for some civil spin-off. Today's economic and technological conditions require R&D to be conducted in a spirit of duality, asking: Does what I seek exist elsewhere? Can what I discover be used elsewhere?[31]

The willingness to place defence interests unashamedly in the forefront of wider national technological development is a theme that was continued, and much developed, in the November 1993 report of the *Groupe de stratégie industrielle* of the French *Commissariat général du plan*.[32]

In sharp contrast to France and Britain, the most powerful economy in Europe, Germany, is also one of the most heavily constrained politically in terms of defence equipment spending, relative to its size. Expenditure on research and technology, as a percentage of the sharply declining procurement budget, has seen a fairly secular rise from about 18 per cent in 1985 to 21 per cent in 1989 and 30 per cent in 1994, but with nevertheless a decline in real terms. Germany has, moreover, had to cope not only with the massive problems of reunification but also, more parochially, with significant changes in the organisation of defence procurement. One aim of this reorganisation is to draw upon civilian technology more intensively in order to reduce costs and improve quality, although in fact it has already been policy since 1985 to draw on civilian-developed technology as far as possible.[33]

Indeed, in contrast to the USA and UK, Germany has not maintained large state-owned defence research establishments. Instead, it makes heavy use of private companies or (to the tune of about one-third of the R&D outlays) other research institutions with primarily civil interests. Industrial contracts for R&D are heavily concentrated in a few firms. These firms themselves have been engaged in extensive restructuring, both nationally (especially with the fusion of Deutsche Aerospace into the Daimler-Benz group) and internationally, and in downsizing. They are on the whole less militarily dependent than their British or French counterparts, and in that sense much more like Japanese defence producers.

Among the smaller defence spenders of western Europe, where the volume of defence R&D is already quite limited, there has also tended to be implicit if not explicit attention to the cultivation of dual-use approaches, which now are being reinforced. Thus in Sweden new dual-use networks are being created to link civil and military R&D organisations.[34]

Overall, therefore, there is evidence of a greater emphasis on dual-use technologies throughout western Europe. It bears adding that, beyond the advantages already mentioned, a stronger emphasis on dual-use technologies would also help with another problem which faces defence planners, namely, how to establish a '*reconstitution capacity*' for defence equipment. The issue here is partly the classical one of having a 'surge capacity', that is, a capacity to expand production of certain items rapidly in a time of emergency. But the concept of reconstitution goes deeper. It acknowledges the point that the ability to design, as well as to build, complex technological systems takes time. This time varies from one type of system to another. The skills needed for design and for project management, as is so often argued when a project is faced with cancellation, are difficult to reassemble once dispersed. Defence ministries on both sides of the Atlantic are known to be considering this problem, trying to assess the time scale on which various design and production capacities could be re-established, and to estimate, on certain defence and foreign policy assumptions, what facilities they should aim to keep in existence under their own control, what they could expect to be able to draw on from allies, and what might be available from the general national or global technology and industrial base. It seems probable that a shift in the direction of dual-use technologies, by rooting defence activities within a more general technology base, could help with this issue also. .

Britain and the European defence market

Debate about these questions can no longer take place in a purely national framework. For Britain, the debate increasingly has a European dimension, which we address here in terms of the industrial and technological levels only.

As has been well recorded,[35] a number of steps have been taken by western European governments to rationalise the defence market, particularly through the 1988 Action Plan of the Independent European Programme Group (IEPG—now the Western European Armaments Group, WEAG). This Plan, with its proposals for movement towards a more open market in armaments, also contains provisions to build up the defence industries of the countries on the geographical periphery of NATO Europe. A cooperative programme of defence research (EUCLID) has been set in place. More specific steps towards reciprocal opening of the British and French markets have also been taken.

These market-opening steps have, however, been extremely modest, both in intent and outcome. The IEPG's Action Plan lacks any legal basis, being essentially only a 'gentlemen's agreement' between governments, backed up by requirements to publicise widely the availability of defence contracts, and to supply, if challenged, the reasons for the failure of losing firms. These, of course, must not include any national protectionist reasons. A critical shortcoming has been the continued application of Article 223 of the Treaty of Rome, which has enabled governments to block the application of Community competition and monopoly regulations to firms or activities deemed to be essential to national security, and hence to maintain the special separate status of the defence sector. Article 223 reads:

> any Member State may take such measures as it considers necessary for the protection of the essential interests of its security which are connected with the production of or trade in arms, munitions and war material; such measures shall not adversely affect the conditions of competition in the common market regarding products which are not intended for specifically military purposes.

The threat of such action has in itself served as a powerful deterrent to intervention by the Commission. Cases of actual invocation of the article have been rare, though one such was its use by the British government in 1994 over the bids by British Aerospace and GEC for Vickers Shipbuilding and Engineering Limited. Although the Commission proposed the abrogation of Article 223 under the Maastricht Treaty, this step was unacceptable to key governments, including the British and French. These governments wish to keep the Commission out of defence (and defence industrial) affairs, though thereafter their paths diverge: Britain seeks full-scale international competition, as well as keeping the lines to the USA open; France, on the other hand, advocates a '*préférence européenne*' in defence procurement, arguing that open competition is a dangerous distraction from the key task of safeguarding an independent European defence industrial base against the US challenge.

As described in greater detail in Chapter 7, the question of an independent European defence industrial and technological capacity has taken a further turn following the Maastricht Treaty commitment to move towards a common foreign and security policy (CFSP), raising the prospect of 'the eventual framing of a common defence policy, which might in time lead to a common defence'. The same article of the Treaty (article J.4) called on the Western European Union 'to elaborate and implement decisions and actions of the Union which shall have defence implications'. And the Treaty stipulated that the Council of Ministers should report to the European Council in 1996 on progress with the common foreign and security policy, with a view to formulating the next steps, while in 1998 the WEU aims to complete the planned revision of its 1948 Brussels Treaty, which will present an occasion to re-evaluate the EC–WEU relationship.

On the question of how to develop the CFSP, an advisory group to External Relations Commissioner, Hans van den Broek, published an interim report in December 1994 which identified the need for a substantial measure of European autonomy in the defence sector if the CFSP is to have any real meaning. Autonomy, the group argued, is required not only for defence production, but also for having access to the necessary expertise and intelligence information to be able to make independent European assessments of potential military threats. At the same time, the expert group recognised the dangers of continuing with the present mode of organisation of defence industrial and technological capacity.[36]

The group sought greater coherence between European military and civil scientific and technological programmes, through some form of linkage between the institutions concerned with military programmes (notably the Western European Armaments Group, the embryonic Western European Armaments Agency, and the Commission), as well as through market integrating steps. The group also called for a study into the possibility of setting up a standing committee of WEU Chiefs of Staff, charged with harmonising medium- and long-term requirements, and another study into the possible role of the Community in managing a coherent process of market opening and restructuring of the arms industries. Against this background, it further suggests, joint studies could be pursued between the WEU and the Commission together, where appropriate, with the European Space Agency, into R&D programmes with a high dual-use content, such as satellite launchers, observation and communications satellites, data transmission, avionics, and a range of other subjects.[37]

In 1995 neither of the two main parties in Britain appeared willing to grant the EU that kind of autonomy, but political leaders should be aware that at the lower level of European research establishments, notably in the aeronautical field, practical steps towards convergence of dual-use research activities are already being taken. These steps began under the auspices of GARTEUR (Group for Aeronautical Research and Technology in Europe). GARTEUR was formed in 1973 by representatives of the government departments responsible for aeronautical research in France, Germany and the UK. The Netherlands, Sweden, Spain and Italy later also joined. Initially GARTEUR supplied an information exchange service, but it soon progressed to organising limited cooperative programmes. It later began to promote and coordinate joint activities where collaboration was seen to be mutually beneficial to the participating nations. In addition to the intergovernmental structure, GARTEUR has also developed a 'shadow' structure comprising leading firms, with which it interacts strongly.[38]

This grouping provided a setting for discussions between the aeronautical research establishments of the seven member states (the 'RE7' talks) about future collaboration. From this network emerged the 'Joint Position on the Future Role of the Aeronautical Research Establishments in Europe' (including, therefore, the British DRA) issued in January 1994 by the seven leading public aeronautical establishments of Europe, and which came into operation in late 1994. Of the seven 'editors' of the RE7 document, five were their nation's representatives on the GARTEUR Council; the two others were from Spain and Italy, and presumably were the link people between GARTEUR and those two countries which do not enjoy full membership. The document envisages that 'A federative process starting with an association, and to be followed by a partnership, should finally lead to a "Union of National Aeronautical Research Establishments in Europe"'. It proposes progressive integration of programmes and facilities in the civil sector, with EU support, but notes that 'A potentially much larger field of coordinated RE [Research Establishment] involvement would be opened up by an established European cooperation in military aeronautics'. It further envisages the possible establishment, in time, of common defence research strategy, possibly resulting in the establishment of a European Defence Research Agency.

Plans are well underway for proposals on substantial rationalisation of European wind tunnels. It remains to be seen how far and how fast these, and other, plans will evolve. What is clear, however, is that there now exists an active policy network representing the interests of government research establishments and firms both to member governments and to the Commission, and with a firm eye in respect of the Commission on lobbying for financial support for aeronautical research in Europe, military as well as civil.

In 1994 a Franco-German armaments agency was launched, to be concerned initially with the management of joint French/German defence equipment projects. After objections by other states that this idea appeared to cut across that for a WEU armaments agency, France and Germany quickly offered to put their agency under the aegis of the WEU, and to represent it as the embryo of a WEU agency, arguing at the same time that they had particular reasons for wanting to get started and not wait for agreement by the other member states.[39]

At the November 1994 Noordwijk meeting of the WEU Council of Ministers, various additional steps in the field of armaments cooperation were taken. Specifically, an informal group of government experts of WEU/WEAG and EU member states was set up to study options for a European armaments policy. The ministers also 'took note' of further work on the idea of an armaments agency, particularly of the Franco-German initiative, and agreed that further discussions would continue, while also 'recognising that conditions do not currently exist for the creation of an agency conducting the full range of procurement activities on behalf of member nations'.[40]

The more ambitious plans for the agency having thus been laid to rest, Britain has now begun actively to consider participation in it, even raising the possibility that management of such major projects as Eurofighter 2000, the Horizon frigate programme, and a new European armoured car could pass to the agency. Earlier in the same month, Prime Minister John Major also called for new defence arrangements for Europe, based on a reinforced partnership between the WEU and the EU, but still rooted firmly in NATO, and firmly controlled at the intergovernmental level.

These developments represent a marked change from the British position of the past several years. The rhetoric, at least, of British politicians and officials on these and related questions has in the past been more muted than that of some of their continental counterparts. The sense that, despite the difficulties, European-level solutions must be found, has seemed to be more strongly held elsewhere in Europe. Britain, on the other hand, both in government and industry, has sought to keep firmly open the link to the United States, from which the UK clearly benefits in the defence field more than all other European states.

It is therefore significant that the British government has begun to engage more publicly and actively with these questions. How far progress will be made, which is to say, how far common visions can be developed and agreed, remains to be seen, not least given the suspicion which many Continental government and company officials hold towards Britain's European intentions. One does not have to take a position on the ultimate wisdom of the different views that are in play to observe that we are at one of those critical junctures from which could emerge a new level of European integration, this time relating to its defence and security identity. It is important, therefore, that Britain be fully engaged in the debate, and that whatever position it adopts should be chosen deliberately, with full consciousness of the consequences for maintaining British defence industrial and technological capabilities.

9 REASSURING CENTRAL EUROPE

Jane M.O. Sharp

This chapter argues that it is in the interest of Britain and her west European allies to extend the benefits of the western security community to the emerging democracies in central Europe. As soon as each state meets the relevant criteria, membership in the North Atlantic Treaty Organisation (NATO) and the European Union (EU) should be offered to Poland, the Czech Republic, Hungary, Romania, Bulgaria, and Slovakia. In the longer term the three Baltic states should also be brought into the EU, but bringing them into NATO will be problematic until the status of the Russian Oblast of Kaliningrad has been clarified.

During the Cold War western Europe enjoyed remarkable economic prosperity, political stability and military security compared with the central European countries. Marshall Aid revived and nurtured war-torn economies while NATO guarantees protected the political and territorial integrity of the western democracies. By contrast, thanks to the spheres-of-influence deal cooked up by Stalin and Churchill in Moscow in October 1944—and sealed in February 1945 at Yalta—the USSR maintained (in most cases installed) repressive Communist governments in Poland, Hungary, Czechoslovakia, Romania, Bulgaria and the eastern half of Germany. In the late 1940s Stalin signed defence pacts with all six and—following the entry of the Federal Republic of Germany (FRG) into NATO—in 1955 Khrushchev subsumed all these bilateral pacts into the multilateral Warsaw Treaty Organisation (WTO) to legitimise further the presence of Soviet troops in central Europe. Though ostensibly modelled on NATO, WTO was not a parallel structure in any sense. Far from protecting their 'allies' Soviet troops forcibly repressed East Germany in 1953, Hungary in 1956, and Czechoslovakia in 1968.

Soviet troops were not, however, deployed on the territory of all six WTO allies. There were none in Bulgaria after the war, and all

Soviet troops were withdrawn from Romania in 1958. Nor were Soviet troops deployed to Czechoslovakia, which had voted in its own Communist government in the late 1940s, until after the crushing of the Prague Spring of 1968. Soviet troops remained in Czechoslovakia, as they did in Poland, Hungary, and East Germany until the early 1990s. Thus, in 1991 when the Soviet Union and the WTO collapsed, a primary goal of all the former non-Soviet WTO countries was to avoid further domination by Russia. An important task for the western allies in the late 1990s is how to reassure their central European neighbours without provoking a Russia already humiliated by the loss of both its internal and external empires.

This chapter looks first at central European security needs, then at the western response to date and finally at how to right the wrongs of Yalta while also meeting legitimate Ukrainian, Baltic and Russian interests.

Central European security dilemmas

After the Cold War central Europeans felt suspended in a security vacuum between the stability of the western security community represented by NATO and the EU, and the turbulence of the Former Soviet Union (FSU) and the Former Yugoslavia (FYU). Once free of Moscow's grip the central Europeans had a number of theoretical options they could pursue towards new security arrangements. Many hoped that the two-bloc Europe of the Cold War would be replaced by a new pan-European collective security organisation. Others advocated neutrality, on the grounds that Austria, Finland, Sweden and Switzerland all fared better than the WTO countries during the Cold War. It soon became obvious, however, that to be credible (i.e. to maintain the kind of independence from the Soviet Union that Finland enjoyed during the Cold War) neutrality required more robust national defence forces than central European governments could afford.

A third possibility was to seek security in a bilateral relationship with a regional hegemony, but for historical reasons neither of the two major regional powers are attractive. On the contrary, history teaches central Europeans to avoid either being caught in the crossfire or squeezed in the embrace of Russia and Germany.[1] Ideally the smaller countries would seek better relations with both these two giants than each has with the other, but those relations need to be within a wider security framework.

A fourth option was a regional grouping that could stand apart from both NATO and the former Soviet republics that made up the Commonwealth of Independent States (CIS). The history of similar efforts in the 1920s and 1930s did not inspire confidence. Nevertheless, under pressure from the west, in February 1991 Poland, Hungary and Czechoslavkia formed the Visegrad Group, but relations between the three remained more competitive than cooperative.

The fifth possibility was to seek membership in western organisations, specifically the European Union and NATO. Former Polish dissident Adam Michnik, now editor-in-chief of *Gazeta Wyborcza* explains why:[2]

> We believe in the democratic rule of law and the principles of human rights, in the effectiveness of the market economy and the power of open societies. We believe in active social welfare policies and an order based on tolerance. We aspire to join NATO and the EU because we fear being marginalised, trapped in our provincialism. We're afraid of isolation, of a solitary existence on the big wide road between Russia and Germany. We believe that by joining European structures, we will become a force for stabilisation at the heart of the continent.

Of these five options, three were soon dismissed: neutrality, seeking protection from either Russia or Germany, and some form of regional group. The first option, a pan-European security system based on the CSCE was the first choice of most central European governments in the euphoria that greeted the end of the Cold War. But the inability of CSCE to cope with the conflicts that erupted in Former Yugoslavia and the Former Soviet Union soon led central Europeans to adopt the fifth option: to seek a western identity in NATO and the EU.

Rejecting Option 1: Losing confidence in the CSCE

In 1989–90, central European governments sought military security by building the Conference on Security and Cooperation in Europe (CSCE) into a pan-European system of collective security. CSCE was then full of promise, not least because since 1975 CSCE standard setting in human rights empowered reformers throughout central

Europe. Dissidents like Vaclav Havel established Helsinki Watch Committees throughout the region to call governments to account if they failed to meet CSCE standards. In addition, the regime of confidence and security building measures (CSBMs), first enshrined in the 1975 Helsinki Final Act, enhanced transparency and predictability between the militaries of NATO and the WTO.

The inadequacy of CSCE mechanisms to deal with post-Cold War conflicts shattered these illusions in the 1990s. Much hoopla surrounded the CSCE summit meeting in Paris in November 1990, which celebrated the signing of the Treaty on Conventional Forces in Europe (CFE) and produced the Charter of Paris for a New Europe. Despite regular summit conferences thereafter, and a plethora of new structures and mechanisms designed to cope with post-Cold War Europe, the 52-member CSCE proved too unwieldy for effective action.[3] Optimists suggested that, with appropriate investment of resources by the major powers, the CSCE tortoise could still overtake the NATO hare to become the most important organisation in Europe. But central European states noted that CSCE operations were used only to monitor (rather than to curb) the wars erupting in Former Yugoslavia (FYU) and in the Former Soviet Union (FSU).

The best hope for an effective CSCE would have been one led by a smaller concert of the major powers. In the early 1990s, the prime candidates for such a concert were Germany plus the four members of the permanent UN Security Council who are also CSCE participants: Russia, USA, UK and France. Relations between the five were good compared with the recent past, with especially close collaboration between Moscow and Washington to strengthen the nuclear nonproliferation regime. Perhaps most important, for the effective functioning of a collective security system, all five powers had well trained armed forces, although Germany was reluctant to deploy military force overseas, and especially not to those countries that were occupied by German forces during the Second World War.

US abdicates leadership after the Cold War

Such a concert never developed because neither the Bush nor the Clinton administration set a high priority on CSCE. In late 1992, with Bush now a lame-duck President, the manifestly superior economic and military capabilities of the USA, combined with the liberal international rhetoric of President-elect Bill Clinton, suggested that the new Democratic administration would support multilateral action to protect human rights wherever they were abused. Europeans, disappointed by the limited efforts of the EU and the UN since mid-1991 to curb Serb aggression, hoped that Clinton would galvanise an international response in 1993. As President, however, Clinton seemed overwhelmed by the commitment George Bush had made to send troops to Somalia late in 1992, and showed no interest in Europe. In his inaugural speech in January 1993 President Clinton said that the US would act with force 'not just to defend its interests, but when the will and conscience of the international community is defied', but for the next two years he made no effort to lead a collective action (through the UN, NATO or the CSCE) to curb genocide and land-grabbing in the Balkans.

The UN and the EU formed an International Committee on Former Yugoslavia (ICFY) in mid-1992, which developed the Vance–Owen peace plan in 1993. The Clinton administration rejected Vance–Owen, but proposed instead that the UN designate and protect a number of safe areas. The UN Security Council established six such safe areas in June 1993, but the UN mandate was never seriously implemented. In 1994 a five-nation Contact Group (Russia, US, France, UK and Germany) took over the UN–EC diplomatic effort. Only in August 1995, however, after tens of thousands of deaths and hundreds of thousands of displaced persons, and the loss of two safe areas to Serb attacks, did the US begin actively to implement the UN mandate to protect Sarajevo, Tuzla and Gorazde and to support the Contact Group plan. This proposal, which would give 49 per cent of Bosnia to the Serbs and 51 per cent to a Muslim–Croat Federation, may end the bloodshed in the short term, but would in effect codify the genocidal ethnic division of Bosnia agreed by the two presidents, Franjo Tudjman (Croatia) and Slobodan Milosevic (Serbia) in March 1991. As such it would reward the aggressors and punish the victims of the war in Bosnia; hardly the model of democratic pluralism the west sought for post-Communist Europe.

Fear that Russia could fill the European leadership gap

Unfolding since 1991, by 1995 the Bosnian debacle revealed the full horror of a leaderless Europe: when the US, Germany, France and

Britain are at odds, Russia begins to call the shots. Exacerbating the loss of European confidence in CSCE in the early 1990s was Russia's campaign to boost CSCE at the expense of NATO. Russian foreign minister, Andrei Kozyrev regularly issued ambitious shopping lists of tasks for the CSCE, but these were rarely well thought out or discussed beforehand with other world leaders.[4] At the CSCE summit in Budapest in December 1994, for example, Kozyrev tried to gain approval for a CSCE European Security Council. Vaclav Havel had proposed something similar in January 1992 at a CSCE ministerial meeting in Prague. But by 1994 Russia's military interventions into the territory of most of its immediate neighbours, as well as the aggressive tone of its military doctrine published in November 1993, made other CSCE participants wary of any directorate that included Russia, even before the attack on Chechnya.[5]

The Budapest summit failed in other ways too. The conference was renamed the Organisation for Security and Cooperation in Europe (OSCE), but delegates could not find common language with which to issue a statement on the situation in Bosnia. They did, however sign a new Code of Conduct on Politico-Military Security, which obligated CSCE governments to protect civilians during the course of internal security operations. Immediately after returning home from Budapest, Boris Yeltsin ordered the bombing of Chechnya.[6] Central Europeans were doubly shocked: first by Russia's blatant contravention of the new Code of Conduct and second by the passive western reaction, for example in not invoking CSCE mechanisms to call Russia to account. Eventually, in April 1995 an OSCE mission was established in Chechnya, but fighting continued through the summer despite various Russian claims that military operations had ceased. The EU suspended trade talks with Russia in January 1995, but in Cannes in June 1995, EU leaders judged that progress towards a political settlement in Chechnya was sufficient to sign a new trade agreement.[7]

To the small central European powers it thus appeared that neither Russia nor any of the EU powers took their OSCE obligations sufficiently seriously and the hopes they vested in CSCE at the end of the Cold War had dissipated by 1995. Few anticipated that the OSCE would ever provide military security guarantees. The most anyone expected was continued oversight of human rights standards and increased attention to conflict prevention and early warning capabilities. The most effective work under OSCE auspices is being carried out by observer and fact-finding missions sent to the CIS countries, and by the former Dutch foreign minister Max van der Stoel, who as CSCE High Commissioner for National Minorities (since December 1992) effectively defused potentially explosive situations between minorities in the Baltic states, Slovakia, Romania, Albania and Macedonia. These are important tasks, but they do not substitute for an effective system of collective defence.

Option V: Joining western institutions

For the central Europeans, joining NATO took precedence over joining EU for three reasons. First, despite the weak leadership of President Clinton, the former Warsaw Pact states still see the transatlantic link with the US as the critical element for security in Europe. In their view, only the US could balance all the potential challenges posed by Russia and Germany on the continent. Second, having observed the indifference of western governments to the crisis in Former Yugoslavia, central Europeans believe that NATO will only protect states inside the alliance, not those in limbo between NATO and the CIS. Czech prime minister Vaclav Klaus added a third reason in a speech to the European Institute in Washington DC in May 1995, namely: 'We need American ideology, American dynamism, free American spirit to counterveil the sometimes opposite tendencies in Europe these days.'[8]

The transatlantic link was the bedrock of European security through the Cold War, and European strategists have every reason to try to maintain strong ties to US policy-makers. Christoph Bertram, foreign editor of *Die Zeit,* suggests one way that NATO governments could rekindle American interest in Europe would be to elect an American as Secretary General of NATO.[9] But the Clinton administration is reluctant to lead. Indeed on a two-day visit to Germany in July 1994 President Clinton appeared to pass the leadership baton to Chancellor Kohl, referring to Germany as America's most important partner and calling on Germany to take the lead on integrating eastern and western Europe.[10]

This was not well received in central Europe, where most leaders believe the best chance for a stable Europe would be a reassertion of American leadership. But if that is no longer on offer, then Europe's best hope is that Germany will take on more of a leadership role not only in political and economic affairs, but also in matters of military security. The international intervention in Bosnia since

1992 demonstrates that France and Britain cannot muster the necessary capability to settle serious crises on their own. If the Americans are no longer willing to help solve Europe's problems, then Germany must take the lead.[11] Those are the only two options for balancing Russia. No-one seemed ready to face this in 1995, however, least of all in Germany. After Hans-Dietrich Genscher's clumsy campaign in 1991 to force the recognition of Croatia on the other EU partners, the Kohl government took little further initiative in FYU and was the most passive member of the five-nation Contact Group established to resolve the crisis in the Spring of 1994.

Western interest in NATO and EU enlargement

The central European governments are all convinced they want to belong to NATO and the EU, but the choice does not rest with them as much as with the members of the two organisations. The western case for enlargement rests largely on the maxim that prevention is better (and less expensive) than cure. Two world wars began earlier this century because of instability in central Europe. It is thus a vital western interest to stabilise the region as soon as possible. Otherwise the western security community may well have to engage in another widespread conflict. Not taking these countries into the EU exposes them to the kind of internal conflicts that stem from poverty and instability, and not taking them into NATO leaves them vulnerable to outside threats, whether from a revanchist Russia or from nationalist rogue states like Milosevic's Serbia and Tudjman's Croatia.

Nevertheless, in the years since the Cold War ended, only Germany consistently championed the enlargement of western institutions.[12] Britain and the other western allies replaced Stalin's Iron Curtain with a curtain of indifference, manifest in the initial efforts to contain rather than resolve the conflicts in Former Yugoslavia, in the barriers maintained against imports from the east, and in the cool response to applications to join the EU and NATO. In central Europe it looked as though the western goal posts were constantly being pushed back or, as one Polish official suggested, 'the west keeps bringing us nearer the soup, but giving us a shorter spoon.'

Even in those western countries which agreed in principle in the early 1990s that NATO and EU should expand eastward, few leaders took their case to their publics to prepare the ground for ratification of subsequent enlargement agreements. This suggested to central European governments that western governments were half-hearted about taking in new members, as indeed many of them were. Disagreements persist about who should join which organisation when, whether NATO and EU enlargement should proceed in parallel or independently, and whether states join as a bloc, or singly as and when they qualify for membership

Who should join what, when?

During 1995, Poland, Hungary and the Czech Republic, which perceived themselves at the head of the queue to join western institutions, were concerned that others, less advanced in their economic and political reforms, would hold them back.[13] But there is no reason why both EU and NATO should not enlarge piecemeal as they have done since their inception. Far from drawing new lines in Europe, the process of enlargement should inspire those in the queue to work harder at reforms. Some aspirants are readier for NATO, others for the EU, and should be treated accordingly.

Hungary and the Czech Republic are obviously both candidates for early entry to the EU. Hungary was the first of the WTO states to experiment with the market economy, as early as the 1960s, and is well on the way to completing the adjustments to qualify for EU membership. The post-Communist government began drafting the framework legislation needed for EU membership in the 1980s and aims to complete the harmonising of Hungarian to EU law by the year 2000.[14] The Czechs started their reforms later but from a much healthier economic base. Czechs too have restructured with EU membership in mind.

Both Hungary and the Czech Republic, however, have been slow to modernise their military establishments. Nor do either have large territories to buffer NATO's eastern border. Thus, while not likely to be as much of an economic drain on the EU as some of their neighbours, Hungary and the Czech Republic could be seen as consumers rather than producers of security, and might not qualify for the first intake of new NATO members.

For Poland, with more territory and a far more impressive army than either Hungary or the Czech Republic, the situation is reversed. Poland is everyone's favourite candidate for early admission to

NATO, despite the unseemly political control that, as President, Lech Walesa sometimes tried to exert over the armed forces. Britain started negotiations to lease Polish training grounds for British armoured units in early 1995.[15] German leaders emphasise the benefits of a Polish territorial buffer for Germany, as well as the importance for the whole of Europe of stabilising Poland inside the western security community. As German defence minister Volke Ruhe put it: if we don't export stability we shall import instability.

Nevertheless, Poland will be more of a drain on EU funds than either Hungary or the Czech Republic. There is thus a case for giving Poland first shot at NATO, where it can make a real contribution, and will also gain confidence to undertake the restructuring necessary for EU membership, while Hungary and the Czechs should be among the first EU intake of former Communist states. This will, *inter alia*, make any wait for NATO easier to bear.

Enlarging the EU

The EU currently numbers 15 members, but began as the European Economic Community (EEC) in 1957 with only six: Belgium, France, Italy, the Federal Republic of Germany (FRG), Luxembourg and the Netherlands. Britain, Ireland and Denmark joined in 1973; Greece in 1981; Spain and Portugal in 1986; Austria, Finland and Sweden in 1995.

The first generation of agreements between the EC and the former Communist countries were trade and cooperation agreements signed in the late 1980s. In December 1991 in the wake of the Maastricht summit meeting, the EU signed second-generation association agreements with Poland, Hungary and Czechoslovakia. These agreements were designed to help aspirants adjust their legal and economic frameworks to that of the EC Single Market, but did not guarantee EU membership, nor were they accompanied by liberalisation in trading those items the central Europeans most need to export: steel, agricultural products and textiles.[16]

Obstacles to overcome

One of the biggest irritants to the central European countries is the EU's Common Agricultural Policy (CAP) which subsidises the farmers of western Europe and discriminates against agricultural imports. Whereas the newest EU members—Austria, Finland and Sweden—are net contributors to the EU coffers, the central Europeans would be net recipients of EU funds. Brussels could not afford to pay CAP subsidies to farmers in central Europe, so the CAP must be modified if the EU is to expand further. Britain's interest thus lies in persuading France to stop subsiding its less efficient but fiercely militant farming community. A useful quid pro quo might be more Euro-centric defence planning by the UK (see Chapters 5 and 9).

Another set of problems is that of competing interests in the EU's under-developed southern regions and in neighbouring North Africa. Three consecutive EU Presidents from Mediterranean countries—France (first half of 1995) Spain (second half of 1995) and Italy (first half of 1996)—tended to push central European membership in the EU onto the backburner.

Border controls are another irritant. In March 1995 Germany, France, Spain, Portugal, Greece, Italy and the three Benelux countries began to implement the Schengen Agreement which relaxes border controls between the nine parties, but toughens controls between the EU and non-EU countries. This new regime exacerbates the resentment that central European states feel about being kept out of the EU. Before Schengen, Poles and Hungarians travelled relatively freely into neighbouring Germany and Austria. Now, however, Hungarian truckers who used to cross into Austria in a matter of minutes face complicated paperwork and queues that take several hours. Poles also complain that Schengen makes them feel like second-class Europeans and has increased from minutes to hours the time it takes to cross the Neisse river from the Polish town of Zgorzelec into Gorlitz.

Nevertheless, despite these difficulties, absent armed conflict in FYU and FSU, there seems little doubt that central European governments would have placed membership in the EU as their top priority. All believe that exporting more goods to the west is the key to prosperity which is in turn the key to political stability. This judgement seems correct. Although the EU moves slowly, once decisions are made they are hard to overturn and members are much more tightly bound than they are in NATO. This in itself becomes a kind of security guarantee. It is the gradual extension of the concept of economic, social and political integration—which has so successfully pacified old enemies in western Europe since the Second World War—that is the best hope for preventing war among the former Communist states in Europe in the next century.

For the nine central European countries with association agreements with the EU, an added benefit is the Associate Partner status offered by the Western European Union (WEU), a club never likely to include Russia. The problem for the central Europeans is the upheaval required in their legal and economic infrastructures to qualify for full EU membership. They also resent what they feel is a lukewarm response from the current EU states. From Warsaw, Prague and Budapest the EU looks like little more than an expression of vested interests of its members. Relative to the EU therefore, NATO looks to have the less forbidding entry requirements.

Enlarging NATO

Like the EU, NATO also expanded piecemeal. Belgium, Britain, Canada, Denmark, France, Iceland, Italy, Luxembourg, Netherlands, Norway, Portugal and the USA concluded the North Atlantic Treaty in 1949. Greece and Turkey joined in 1952, the Federal Republic of Germany in 1955, and Spain in 1982. Soon after the WTO collapsed, in October 1991, the three Visegrad states (Poland, Hungary and Czechoslovakia) announced that, given the dangers posed by the disintegration of Yugoslavia and the likely breakup of the USSR after the August 1991 coup against Gorbachev, they wanted to join NATO.[17] Since then Bulgaria, Romania and the three Baltic states have also expressed a desire to join.

Except from Bonn, where the government stresses the importance of promoting stability in central Europe, the response from most NATO capitals has been cool. Governments have three main worries: that decision-making will become less manageable with more than sixteen; that extending security guarantees might entangle the NATO countries in conflicts not of their own choosing; and that enlargement will provoke Russia.

It is perhaps not surprising that North Americans and Europeans have not always been in tune on NATO enlargement as they have always differed on the purpose of the alliance. West Europeans see the threefold benefits of NATO as containing Russia while engaging both the US and Germany in the western security community. North Americans, however, see NATO almost exclusively in terms of containing, and if necessary countering, the Soviet threat. These differences did not matter during the Cold War years. All the allies wanted a strong NATO, wanted the US to lead, and were prepared to live with a sometimes difficult burden-sharing relationship. After the Cold War, some Americans argued there was little need for as strong a US commitment to Europe, that the main goal was to bring a democratic Russia into a pan-European security system. This Russia First view dominated the first 18 months of the Clinton administration, delaying important decisions on enlargement. Henry Kissinger even accused President Clinton of buying the Russian argument that NATO was hostile to Russia.[18]

The North Atlantic Cooperation Council (NACC)

NATO's first overture to the former Communist states came in late 1991, with invitations to all the former WTO allies, including all the former Soviet republics, to join the sixteen NATO states in a North Atlantic Cooperation Council (NACC). NACC had two main goals: to curb the appetites of former WTO allies for full NATO membership and to rationalise a NATO *droit de regard* over the reorganisation of the former Soviet military forces among the newly independent republics. NATO also promised the former WTO powers consultation and cooperation on security issues including defence planning, arms control concepts, civil-military relations, coordination of air traffic and the conversion of defence to civilian production.

NACC was disappointing to the leaders of Poland, Hungary, the Czech and Slovak Republics, who all believed they deserved preferential treatment by NATO on the grounds that they were manifestly closer to achieving western-style democracies than their former WTO partners. On the positive side, however, NACC facilitated the allocation of former Soviet conventional equipment among the former Soviet republics, a necessary precondition for ratification of the 1990 Treaty on Conventional Forces in Europe (CFE). NACC also provided a useful forum in which the western powers and the four republics which inherited Soviet nuclear weapons on their territory (Russia, Kazakhstan, Ukraine and Belarus) could discuss how these states could be denuclearised and incorporated in the nuclear non proliferation regime.

Central European states seeking full membership in NATO were encouraged in September 1993 when NATO Secretary General, Manfred Woerner, noted that NATO was not a closed shop and that nothing in the 1949 North Atlantic Treaty precludes adding new members.[19] In October 1993, after the success of Zhirinovsky in the Russian elections, and the increasingly anti-western line adopted by foreign minister Andrei Kozyrev, many officials in the foreign and defence ministries of western Europe also assumed the January 1994 NATO summit would move the Visegrad group closer to full membership.[20]

But there was still no enthusiasm for enlargement in Washington. During 1993, west European officials complained that it was difficult to ascertain Clinton administration policy. In the Russia First camp, Strobe Talbott, Defense Secretary Les Aspin and Warren Christopher all argued against enlargement, whereas National Security Adviser Anthony Lake and Vice President Albert Gore, more sensitive to the needs of Europe, argued in favour. Initially the Russia First camp prevailed, with a second delaying tactic in the form of the Partnership for Peace (PfP), first floated by Secretary of Defense Aspin at a NATO ministerial meeting in Travemunde in October 1993. The other NATO countries approved of PfP in late 1993 and invited all the CSCE countries to participate in January 1994.[21]

Partnership for peace: old wine in new bottles?

Despite their disappointment, the central Europeans worked hard to become good partners in the PfP. Participation came in two stages. In the first stage the North Atlantic Council drew up a Framework Document for prospective partners. Individual partner countries then drafted their own Presentation Document listing the assets they could offer to the PfP. NATO then viewed each Presentation Document on its merits, before concluding a bilateral Individual Partnership Programme (IPP) with each partner country. PfP relationships are like spokes of a wheel the hub of which is NATO of the sixteen.

The formal objectives of PfP, as set out in the Framework Document, replicate those of the NACC: to facilitate transparency in national defence planning and budgetary processes; ensure democratic control over military forces and defence industries; maintain forces capable and ready to participate in UN or CSCE (now OSCE) peacekeeping operations; to develop habits of cooperation with NATO forces and to facilitate joint actions within the NATO integrated command structure. The practical programmes include joint planning and exercise; the right to consult in the event of a partner state being threatened; permanent liaison staff at NATO HQ; a Partnership Coordination Cell (PCC) at SHAPE in Mons; and a planning and review process of the Presentation Documents.[22]

PfP set no specific targets for NATO membership, however, either in terms of criteria that states should meet, or a schedule for taking in new members; and offered no security guarantee. To the central Europeans PfP reflected NATO's acquiescence in a Russian *droit de regard* over its old empire and a Russian veto over NATO membership. These fears were confirmed at the June 1994 meeting of NATO and former WTO foreign ministers in Istanbul, when Russia succeeded in removing language from the final communiqué that referred to east European membership in NATO.[23] Western capitulation to this Russian veto was all the more frightening to the smaller European powers in the wake of western appeasement of Serbian aggression since 1991.

Beyond PfP?

Poland, the most energetic in pursuit of NATO membership. was initially the most upset by PfP. In December 1993 foreign minister Andrzej Olechowski visited Washington, not least to ginger up the Polish-American lobby to remind Bill Clinton that he would need to win both Pennsylvania and Illinois if he had a hope of re-election in 1996. The pro-enlargement camp in the US received another boost early in 1994 when Richard Holbrooke returned to Washington from Bonn (where he served as US Ambassador) to take up a new post as Assistant Secretary of State for European Affairs. Holbrooke was more sensitive to the needs of Germany and central Europe than most senior Clinton administration officials, and pushed hard for NATO enlargement primarily to promote stability throughout Europe. Besides, the Republican-controlled congress wanted to take in new NATO members to send a tough message to Moscow. Thus, across the political spectrum, pressure built up through 1994 for a decision on enlargement. Even Strobe Talbott, the original 'Russia Firster', now endorsed enlargement.[24] The results were meagre, however, and represented yet another delay for the central Europeans. In December NATO ministers initiated a study on enlargement but with no specific timetable.

Making NATO enlargement palatable to those left out: Ukraine, the Baltic states and Russia

Some critics of NATO enlargement fear that it will draw a new line in Europe leaving those states between the expanded NATO and Russia vulnerable to pressure from Moscow. If any state was drawing new lines in Europe in the early 1990s, however, it was Russia, as Moscow sought unambiguously to reassert control over the CIS states. Some CIS states, Belarus for example, appeared relatively content to draw closer to Russia, so the worry about states being left out of the first NATO intake applies primarily to Ukraine and the three Baltic states.

Assuring Ukraine

NATO should obviously not expand in such a way as to give Moscow any pretext for absorbing Ukraine into the Russian orbit. This implies both carrots and sticks for Ukraine: gentle sticks in the form of encouragement to complete the denuclearisation process as rapidly as possible and, if necessary, restraints on behaviour that might seem provocative to Russia—undermining Russian minority rights for example. Carrots should be in the form of economic aid with special efforts to open up trade and offer technical assistance, especially to help make Ukraine as independent of Russia as possible in energy resources. Western governments should make clear to the Kremlin that any cooperative agreements with Russia are conditional on the independence of Ukraine, the model for which should be Finnish neutrality during the Cold War. This will require a substantial level of investment from all the western democracies.

Reassuring the Baltic states

Since achieving independence in 1991 Estonia, Latvia and Lithuania have joined the Council of Europe, signed up to NATO's PfP, become Associate Partners of the WEU and are on track for membership in the EU. Lacking the military power to defend themselves, all three Baltic states are also anxious to join NATO. For NATO, however, the Baltics would be far more difficult to absorb than the central Europeans, because of the large ethnic Russian populations in Latvia and Estonia and the position of Lithuania sandwiched between the heavily militarised Russian oblast of Kaliningrad and the rest of Russia. Russian elites seem less upset by the prospect of their former satellites joining the EU and the WEU than NATO and, at least in the short to medium term, this seems the best option for the Baltic states. None of Baltic states are close to being ready to join either institution, but they should probably be discouraged from aspiring to NATO until Kaliningrad has been demilitarised.

Meanwhile, in addition to the NACC and PfP projects, Britain and the Nordic NATO countries can reassure the Baltic states through various cooperative military ventures. A good example is the Baltic peacekeeping battalion (BALTAP), which includes the UK and Denmark, and served with UNPROFOR in Bosnia.[25] As neighbours, former neutrals Finland and Sweden also have special responsibilities to Estonia, Latvia and Lithuania. Former Swedish prime minister Carl Bildt said in 1994 that Russian behaviour to the Baltics would be a litmus test of the direction of Russian foreign policy and suggested that Sweden should drop its Cold War neutrality in favour of some form of security guarantee to the Baltics. This is a controversial notion in Sweden where neutrality is a political icon.[26] But Sweden enjoyed a free ride in security terms through the Cold War, and has a responsibility now to pitch in with aid, trade and technical assistance to reassure the Baltic states that they belong to a well-integrated network of democracies in the Nordic area.

Russian attitudes towards NATO

The Russians have always been ambivalent about NATO and about the US role in Europe. During the Cold War, hard-line Soviet officials like Andrei Gromyko often said the US had no business in Europe, but Soviet delegates to the NATO—WTO negotiations on mutual and balanced force reductions (MBFR)—in Vienna from 1973 to 1989—said that NATO, and especially the deployment of US troops in Germany, were important stabilisers for the whole of Europe.[27] In 1990, Gorbachev initially opposed German unification, then said unification was acceptable as long as the new Germany was not in NATO. Eventually Gorbachev withdrew his opposition altogether as Germany pledged unilaterally to reduce the armed forces of the combined Germany to 370,000, which was less than former FRG levels. East Germany was always portrayed as the main Soviet prize of the second world, so the loss to NATO was especially hard for the Russian military.

Even harder for Russia to accept was that all the other former non-Soviet WTO members then wanted to join NATO. Nevertheless to argue as Yeltsin and Kozyrev did in 1994—95 that enlargement is inherently anti-Russian is to ignore NATO's increasingly conciliatory policy during the 1990s towards all the former WTO states, including all the former Soviet republics. The NATO countries repeatedly demonstrated a desire to engage Russia in a pan-European security dialogue, and invested substantial time, effort, personnel and funds to do so in a number of military exchange visits, education and training schemes, joint inspections, and denuclearisation programmes as well as all the individual NACC and PfP programmes. If Russia isolates itself, it will be of Moscow's own choosing.

When Russia and the other former Soviet republics signed up to the NACC in early 1992 they appeared relatively relaxed about NATO enlargement. The following year, in August 1993, on a visit to Warsaw, Yeltsin issued a joint statement with President Lech Walesa, accepting Poland's bid to join NATO. Senior military officers challenged Yeltsin on NATO enlargement on his return to Moscow and on 15 September Yeltsin wrote to the French, German, British and American heads of state to assert that the 1990 Treaty on the Final Settlement with Respect to Germany (which prohibits the stationing of foreign forces in the five eastern *länder* of a unified Germany) 'rules out any possibility of a NATO expansion eastward'.[28] It does no such thing of course. On the contrary, given the grim legacy of the WTO for the central European states, the right of independent sovereign European states to join any alliance of their choice was explicitly provided for in a number of CSCE documents after the WTO collapsed.[29]

Not all segments of the Russian elites opposed NATO enlargement. Sergei Blagavolin, a defence analyst with the Russia's Choice party, believed NATO enlargement would help to stabilise Europe.[30] Most of the leaders of Russia's Choice, including Yegor Gaidar, took the more cautious line that, although they personally saw nothing negative about NATO of the sixteen, enlargement prior to the Russian elections in 1995—96 could strengthen the red-brown (Communist—Fascist) coalition. Alexei Arbatov, a defence specialist and member of the Duma representing Yabloko, said that if Russia invaded Ukraine (or did something equally destabilising) then NATO should obviously expand, but otherwise it was better not to stir up Russian nationalists.

In January 1994 Kozyrev published an article in the *Frankfurter Rundschau* conceding that Russia had no right to stipulate who joined NATO, but insisting on the principle that NATO remain open to any democratic European or Atlantic state that wished to join.[31] Boris Federov, the leader of Russia Forward, also took the line that NATO must not exclude Russia from membership, even suggesting that Russian participation in the alliance could displace the US role of balancing a stronger Germany.[32]

Although Kozyrev threw a tantrum in Brussels in December 1994 when NATO launched its one-year study of membership criteria, in early 1995 several senior officials in the Yeltsin government appeared resigned to NATO enlargement. On 17 February in Warsaw, for example, prime minister Chernomydin repeated that Russia would put no obstacles in the way of Poland joining the alliance.[33] About the same time Yeltsin assured President Gyula Horn that Russia did not oppose Hungary joining NATO. Then, in early March, Kozyrev (in conversations with Warren Christopher) and Georgi Mamedov (in conversations with the FCO in London) began to lay down a set of Russian conditions for NATO expansion.[34] These included no forward-based nuclear weapons; no forward-based NATO troops; no new members before the Russian elections in late 1995; no permanent exclusion of Russia; new NATO members not to join the integrated military command of NATO (like France and Spain); no NATO exercises in the former WTO states without prior consultation with Russia; and revision of CFE to accommodate the changed circumstances.[35]

On 15 March 1995, however, a directive from Yeltsin to the Russian Ministry of Foreign Affairs stopped further discussion of conditionality with reference to NATO enlargement. Thereafter Russian statements on NATO became much tougher.[36] For example, that NATO could only admit new members after conclusion of a new Russia-NATO agreement; that enlargement would sabotage CFE, trigger new alliances with Iran and Iraq, and generate more overt help to the Serbs in Bosnia.

Some Russian reformers despaired that the political classes in Moscow would become hostage to their own rhetoric on NATO enlargement.[37] At worst a continued negative reaction could re-establish the old patterns of confrontation as an excuse to build up the military and to abrogate existing arms control agreements. The best case would be for Russia to accept agreed NATO policy, to re-affirm a new partnership between NATO of the sixteen+ with

Russia, to reaffirm Russia's role as a great power with special interests in Europe and, most important, to begin a pro-active policy of reconciliation with central Europe. The likely middle road will be a continued bargaining for concessions (as with German unification in 1990) and an attempt to repair the damage already done to Russia's reputation through an improved relationship with the NATO countries.[38]

Sweetening the deal

The Study on NATO Enlargement makes some effort to meet Russian concerns. For example. while the study notes that exercises will be held regularly on new members' territory, and that new members will share the benefits and responsibilities of the nuclear guarantee in the same way as all other allies, it also states that the alliance has no *a priori* requirement for the stationing of alliance troops on the territory of new members. [39] Statements made by officials in Brussels and other NATO capitals made clear that no invitations would be issued to specific countries until 1997, that is after the elections in the US and Russian and after the inter-governmental conference (IGC) of the EU.

While Britain and the other NATO governments must make clear to Russia that the security of the states that aspire to join NATO are not chips to be bargained away at some international poker game, there are a number of ways to sweeten enlargement for Moscow. The first is that parallel to taking in new members NATO should enhance its conciliatory outreach towards all the former Soviet republics. This includes building on the network of contacts with parliamentarians and military personnel developed since the end of the Cold War through NACC and PfP as well as the North Atlantic Assembly. NATO should also emphasise the extent to which it has cut back defence budgets, withdrawn nuclear weapons and dismantled heavy forward-based equipment to demonstrate that the alliance is now defensively oriented. NATO is already far more transparent in its budgeting and force planning than any of the CIS countries, but should take whatever opportunities are offered for more openness.

A second sweetener, already under way, is to forge a new treaty-based partnership between Russia and NATO. A preliminary document was initialled in November 1994 which Kozyrev refused to sign as scheduled in December, claiming NATO had reneged on a previous understanding that PfP was an alternative to enlargement. The Russian attack on Chechnya further delayed the process, but Kozyrev eventually signed both the Individual Partnership Programme (IPP) for PfP and the NATO—Russia Document on 31 May 1995 at the NATO ministerial meetings in Noordwijk.[40] The first practical meeting to implement both documents was held in Brussels on 17 July 1995.

A third possible sweetener is dealt with in greater detail in Chapter 14, namely for NATO to take more seriously Russia's repeated request to revise the Treaty on Conventional Forces in Europe, signed by all the NATO and WTO states in November 1990, but which the Russians claim has been overtaken by events.

In addition to the sweeteners, however, there must be a firm and consistent policy of holding Russia to international norms of behaviour. The western democracies owe this to Russia's weaker European neighbours, and to Russian democrats and reformers too. If Russia does not know what western expectations are, its Great Power instincts will constantly test western limits, alienating it further and further from the international community.

For a consistent European policy towards Russia, the next British government must stick close to Germany in terms of future European policy-making, both to build confidence throughout Europe that we are on the side of generously expanding western benefits eastward, but also to nudge Germany into those policy areas where she is currently reluctant to lead. To be able to reassure the former Communist countries the western allies need to establish a set of principles on which to build a stable and secure Europe and begin to apply them consistently. Refusing to tolerate genocide and the changing of borders by force would be a good place to start.

PART 4

BRITAIN IN THE WIDER WORLD

10 DOING BUSINESS WITH THE FORMER SOVIET UNION

Neil Malcolm

The former Soviet Union represents a special problem for Britain, and the rest of western Europe, because of its size, its closeness and its unpredictability. Russia alone is a massive presence on the European continent. It has a population of 150 million, the overwhelming majority living west of the Urals, over one-and-a-half million men under arms, and tens of thousands of nuclear weapons on its territory. It occupies a strategically central position in Eurasia, and dominates its less well-endowed neighbours in the former Soviet space. It maintains military bases in all of them except Azerbaijan and its troops are active in what are described as peacekeeping operations in several parts of the region. Although the economies of the CIS states are currently in a devastated condition, the region's impressive natural resources and human capital could make it a zone of rapid economic growth in the next century.[1]

Immediately after the dissolution of the Soviet Union, unaccustomed restraint on Russia's part in 1991–92 almost caused the west to forget how much disruption it could cause in zones of instability such as central and south-eastern Europe and the Middle East. From 1993 more active Russian diplomacy has reminded us (as it was intended to) of what could happen should the country revert to a strategy of confrontation.

If geography and geopolitics make the former Soviet Union vitally important for western Europe, history and culture make it particularly difficult to understand and deal with. It would be comforting to think that now that Communism has passed we can approach the FSU as if it were a collection of 'normal' states, reacting in a fairly predictable way to international threats and opportunities. Unfortunately this is not the case. Indeed change and uncertainty and loss of bearings in the east have made things more imponderable.

Some analysts suggest that the western democracies take George

Kennan's advice, and 'deal with Moscow not with platitudes, but on the basis of a firm and realistic understanding of respective interests'.[2] It is true that Russian international relations writers and politicians talk a great deal about national interests, and indeed there are economic and security interests of a specific kind which are widely agreed on. However, unlike in the 1940s and 1950s when George Kennan was sending his despatches, in Moscow in the 1990s there is no settled view of more general foreign policy interests and no entrenched basic foreign policy strategy.

All this means that shaping policy towards Russia and its neighbours requires constant attention, sensitivity to special local features, imagination and flexibility. In 1990 and 1991 it could be said that the western alliance rose well to the challenge of the collapse of Soviet power, and did its part in ensuring a remarkably bloodless transition. In the period since then, however, interest has been sporadic and progress has been less impressive. The resulting picture resembles one of the unfinished building sites familiar in the Soviet period—bare foundations, skeletal structures standing here and there, piles of components rusting in the rain.

One difficulty has been to get away from thinking in the old terms about the east and about our own institutional structures. Black-and-white perceptions of a struggle between 'communists' and 'democrats' or of a 'come-back of Russian imperialism' have not always been helpful in understanding Russian foreign policy since 1992, for example. Similarly, while it may be more prudent and practical to adapt existing institutions to suit the new demands of European and world politics than to invent new ones, this requires both an unusual amount of determination by leaders to overcome fear of change in these institutions, and imagination in adapting them.

Another difficulty for western policy-makers has been to sustain a coherent view of eastern realities, and to unite around common priorities and policies. The attitudes of western governments appear more and more to have been coloured by partisan perceptions and particular interests. By 1995 echoes of Cold War rhetoric were sounding around discussion of Russian policy in Washington. In Germany old geopolitical themes were becoming more obtrusive. In France the defensive concern to safeguard EU integrity was uppermost. Meanwhile little progress was made in providing the former Soviet states with a clear picture of their future place in Europe, or with an unambiguous set of guidelines about the kind of behaviour that was required of them. As a result, developments were allowed to drift at a crucial period in post-Soviet domestic politics.

For the United Kingdom the former Soviet Union appears more distant then it does for Germany, say, both in terms of economic interest and in security terms. The advantage of distance is that Britain can be less constrained in its policies towards the East. It can give a political lead at crucial moments, as it did in 1984–85, when the opportunities represented by Gorbachev's rise were first becoming apparent. The United Kingdom also has the assets of substantial knowledge and experience of Russian and CIS affairs, and familiarity with other parts of the world, for example south and east Asia, which are important for Russian foreign policy. Britain's importance in international education and training makes it especially well qualified to take the lead in dealing with Russia's heritage of cultural isolation during the Cold War.

The United States can perform a similar role, but the American medium- and long-term commitment is less certain. If our aim is to try to work towards a common European foreign and security policy, then Eastern Europe, where German interests, for example, are so deeply engaged, is a crucial area to focus on. The problems of working with Russia and its surrounding states could turn out to be a unifying challenge or a disintegrating force.

Perceptions

It is important for western partners to take into account the fluidity and open-endedness of post-Soviet affairs. We must look below the surface to understand how Russian views of the world have been developing over the last decade. Only on that foundation can we make reliable guesses at possible future paths of Russian policy and at the kind of factors which will be important in determining them. This chapter thus begins with a review of recent changes in Russian thinking about international relations, and especially relations with the west. The second main theme will be developments in Russian and CIS politics which are likely to be important for foreign policy. Finally, the chapter considers western policy dilemmas in dealing with the former Soviet Union.

For centuries 'westernism' and 'anti-westernism' have been key elements, at times dominant elements, in Russian politics and foreign policy. On one side there were those who saw the west primarily as a military threat, or a threat to Russian culture and way of life. For others it was a pole of attraction, a social model, and wider contacts with it were essential for economic and technological progress. Drawn into the European orbit, but struggling to defend its independence and separate identity, Russia passed through a series of phases of opening up to the west, accompanied by internal westernising reforms, followed by phases of relative isolation and social conservatism. Under Stalin and his successors the country endured the most nationalist and inward-looking regime that it had had for 100 years.

It is impossible to understand the nature of the westernising impulse which had such dramatic effects in the Soviet Union after 1985 without being aware of the depth of isolation and ignorance of the outside world which preceded it. Stalin's uncompromising anti-westernism fostered an almost unconditional westernism among reformers of the next generation. Gorbachev's New Political Thinking opened the way to a transformation of official doctrine: the rhetoric of international class struggle was replaced by appeals to global, all-human values, imperialist exploitation was redefined as globalisation and interdependence, and Stalinist pan-Europeanism aimed against NATO and EEC consolidation evolved into acceptance of European integration around the western core. As Soviet foreign minister, on signing a trade and economic cooperation treaty with the European Community in December 1989, Eduard Shevardnadze spoke of the pressures for integration 'spreading across the continent in waves' from the west.[3]

After the Second World War, Europe pioneered a regional integration which made war between France and Germany unthinkable. After the Cold War, many westernisers hoped this project could be extended throughout the continent. In the words of deputy foreign minister Vladimir Petrovsky, the new Europe was 'a component and a prototype of a new system of human relations built on the principles of non-violence, solidarity and cooperation'.[4] Although the new Russian foreign minister, Andrei Kozyrev, criticised the New Political Thinkers for their 'idealism', he followed in their tradition, stating in 1992, 'Europe is waiting for us, and we are ready to enter it', and declaring his intention of employing foreign policy as it had been used under Gorbachev, 'as a sort of locomotive for the Union which hauled it into civilization'.[5] Russia, in his words, was 'the missing component of the democratic pole of the northern hemisphere'.[6]

At the NATO summit in London in July 1990 and at the signing of the Charter of Paris in December, western diplomacy worked to sustain this mood. All sides affirmed their commitment to building a new, peaceful European order, and in particular to strengthening the mechanisms of the Conference on Security and Cooperation in Europe. The Conventional Forces in Europe (CFE) treaty was signed, and a programme for military cooperation between NATO and the former members of the Warsaw Treaty Organisation (WTO) led to the founding of the North Atlantic Cooperation Council (NACC) a year later. The USSR was a shareholder in the new European Bank for Reconstruction and Development (EBRD) established in the middle of 1991, and negotiations continued for a new agreement with the EC-based on Article 238 of the Treaty of Rome, which deals with association agreements. Germany played a leading part in the process of building trust, acting as Moscow's advocate in European and western forums and providing the biggest share of the first wave of economic assistance.

During the 1990s, however, Russian hopes of 'joining the club' were repeatedly disappointed. Organisational interests in existing western European institutions were threatened by the prospect of adjusting to admit an entity as large as the former Soviet Union. In the case of NATO there was understandable difficulty in accepting the idea of admitting the former Soviet states that NATO was originally designed to contain. Western states had quite legitimate misgivings about future political stability in the former Soviet space, about its commitment to democratic values, about its capacity for economic reform, and about the costs of providing assistance on a scale that might have any positive impact.

Thus the European Union was prepared to admit central European states to the west of the former Soviet Union, but Russia and the other members of the CIS would remain outside for the foreseeable future. NATO likewise has been willing to contemplate only relations of partnership rather than membership. Russia was invited to the annual Group of Seven summits, but only for the political part of the discussions. And in general Moscow complained that it was not being regularly consulted even on international issues where it had a strong interest, such as making peace in the former Yugoslavia, or redesigning the COCOM export control system.

The failure of the CSCE to develop into an effective new pan-European security organisation, despite the rhetoric of 1990, and the continuing vitality of NATO were particularly difficult to reconcile with the Russian westernist conception of the future of Europe. The idea of NATO expansion was bound to be a difficult one to sell to certain parts of the Russian elite. Internal political changes which led to the military becoming much more influential, and to a competition in 'patriotism' among Russian politicians made it even more awkward. Changes in Russian thinking about international relations in general, and a swing away from the idealism of New Political Thinking, were also important. Russian commentators began to point to the failure of international mediation efforts to stop the fighting in Croatia and Bosnia and compared it to the success of the decisive, virtually unilateral military action by the United States against Iraq in 1990.

The conflicts which sprang up around the southern periphery of the former Soviet Union provided another reminder of the continuing importance of the organised use of force in international affairs. Critics of Kozyrev's westernism described themselves as 'realists' and called on him to fight harder for Russia's national interests in a competitive world. They were able to contrast the disappointingly small improvements won in the civil rights of Russians in the Baltic states using appeals to institutions such as the Council of Europe with the effective snuffing out by the Russian army of civil wars in Moldovan Transdnestria and Georgian Abkhazia.[7] The independent-mindedness of new states like Ukraine and Azerbaijan came as a shock to Moscow, and encouraged worries about security threats on Russia's southern flank. This also fed the resurgence of geopolitical, balance-of-power thinking.

By the second half of 1992, the implications of the break-up of the Soviet Union were having an important effect on Russian views of the world. Many who were happy to see the cancellation of the Soviet Union's global commitments and who were prepared to accept the 'loss' of eastern Europe found it impossible to come to terms with the shrinking of Russia to its sixteenth-century borders. One effect was an upsurge of 'Eurasian' thinking among foreign policy experts. They argued that the Europeanism and Atlanticism of 1989–91 had led to a dangerous neglect of problems in Russia's own backyard. The consequences were already evident, they claimed, in economic disruption, threats to the safety of Russians living in the new 'near abroad', and the prospect of political inroads being made by countries like Turkey and Iran. They called on the government to redress the balance and pay more attention to the east. This went along with calls to modify internal reform plans, on the grounds that radical westernization was impossible in Russian conditions.

In the economic sphere there were increasing complaints that Russia was being excluded from world markets, especially those for high-technology goods. In 1992–93 political capital was made by the opposition when the United States persuaded Moscow to cancel the transfer of cryogenic rocket-engine technology to India. Many Russians were disappointed in the relatively small scale of western economic assistance, in the inappropriate conditions attached to foreign aid. They complained that most of the benefits were channelled to corrupt Russian officials and criminals and to producers and consulting firms in the west.

In retrospect it is striking how smoothly the dismantling of the Soviet empire came about, but it would be a mistake to assume that severe problems of post-imperial mental adjustment have been miraculously avoided. Fairly widespread among the elite as a whole (and particularly acute among the military) are straightforward disorientation and resentment caused by the decline in Russia's international standing, the sudden influx of western culture and the undermining of traditional values. Attitudes among more westernised groups are more complicated. After the hopes (however unrealistic) of 1989–91 there was a sharp sense of having been let down by the west. The nationalistic tone adopted even by reformist politicians and experts in Moscow from 1993 onwards was not just an effect of adapting to domestic political pressures. It also reflected a contradictory mixture of feelings which for the west represent both a danger and an opportunity.

Politics

Gorbachev's perestroika and the radical changes which followed it were accompanied by a fall in the fortunes of important occupational groups—Party officials, the security police, the military, employees in the defence industry. Setbacks abroad and failure to be welcomed unambiguously into the western community strengthened the opponents of reform at the same time as they demoralised its supporters. However, it was internal failures which had the main effect on the political atmosphere: the abrupt decline in production and threatening industrial collapse, spreading crime, corruption and disorder, what seemed a dangerous weakening of respect for law and the authority of the state.

Yeltsin and his reformist ministers were thrown onto the defensive as early as the spring of 1992. Faced with more and more bitter opposition in the parliament, Yeltsin set out to broaden his support base, making compromises with the political 'centre'—industrial managers and military leaders who were ready to give him their support on condition that the pace of reform was reduced and that efforts were made to restore a single economic and strategic space in the CIS. The attempt to apply economic shock therapy was abandoned, and in early 1993 Yegor Gaidar was replaced as prime minister by Gazprom chief executive Viktor Chernomyrdin.[8] The Ministry of Defence came to wield unusual influence over aspects of foreign policy normally thought of as political, especially in the CIS region. One Russian journalist commented that whereas the state had adopted dinner-jacketed (Foreign-Ministry) policies in the wider world, it preferred a flak-jacketed policy in the 'near abroad'.[9] Worse yet, the Kremlin appeared either unable or unwilling to discipline commanders who took sides in local conflicts instead of carrying out an impartial peacekeeping role.

It would be incorrect to describe the 'centrists' as anti-western. They were pragmatic individuals who recognised that confrontation had led the USSR into a cul-de-sac. But many of them had authoritarian, statist reflexes, and they found it difficult to accept the full sovereignty of the other former Soviet states. Especially after parliament was forcibly dispersed in October 1993, Yeltsin came to depend more on his image as a strong leader and energetic defender of Russia's interests abroad than on his reputation as a democrat. The political weight of the armed forces and the security police increased.

In the December 1993 elections support for liberal westernisers slumped, and Zhirinovsky's extreme-right Liberal Democratic Party became the second largest in the lower house, after a particularly good showing among army officers. Yeltsin and Kozyrev then began to sound even more 'patriotic' in their foreign-policy statements. In negotiations with the west they complained of the constraints placed on their freedom of action by nationalist currents in public feeling. There is some reason not to take this at face value. Zhirinovsky gained votes mainly because of disillusionment with socialism and liberalism and the popularity of his hard-line 'law and order' proposals, not because of his extravagant line in foreign affairs.

Public opinion at large is little concerned with the world outside the boundaries of the CIS. It regards the west with a degree of suspicion, but sees no alternative to reaching an accommodation with it.

Russian opinion shifted in a more assertive direction, but mainly among the political elite. Simultaneously more nationalist elements among the elite gained in political influence. These two factors made Yeltsin readier to ignore criticism from liberals and from the west, and to push ahead with adventures such as the invasion of Chechnya at the end of 1994.

Where Yeltsin must take wider public opinion into account is concerning relations with the other parts of the former Soviet Union. The frontiers of the fifteen Soviet Union republics were drawn with little regard for ethnic/linguistic boundaries and sometimes cut deliberately across them. The RSFSR, now the Russian Federation, thus excludes large Russian-speaking areas. In the mental universe of most Russians, most of the rest of the former Soviet Union is still part of 'Russia', and what happens in the near abroad seems to belong, as the term implies, to an intermediate zone between domestic and foreign affairs. The evidence of opinion polls is that even the most liberal-minded Russians have so far made little progress towards accepting the breakup of the Soviet Union and the sovereign rights of the other CIS states. Political parties universally affirm the urgent need to 'protect' the 25 million Russian speakers and to guard Russian interests in general in the near abroad.[10]

Conversely, large parts of the population (not just those who identify themselves as Russians) in many of the newly independent states find it difficult to come to terms with the idea of total separation from Russia. For example in Belarus and in central Asia, independence arrived largely unwanted and unexpected. In Ukraine and Kazakhstan, between a quarter and a half of the population describe themselves as Russians, and many more have Russian as their first or only language. In most cases, disruption of economic ties with Russia had extremely damaging results.

All this gives rise to centripetal forces which are often ignored by outsiders and which complicate any project of encouraging 'geopolitical pluralism' in the former Soviet Union.[11] Even the Baltic states, which enjoyed independence between the First and Second World Wars, and which lie outside the CIS, have been areas of Russian influence since the eighteenth century, and still contain millions of Russians. For us the right of these nations to make the decision to join the EU and even NATO seems self-evident. For most

Russians, however, the idea of the Estonian border, say, becoming a new strategic dividing line between east and west seems unnatural and threatening.

The general domestic political context of Russian foreign policy in 1995 thus appeared quite different from 1992, when Yeltsin was able to pose as a champion of democracy and friendly relations with the west, besieged by a violently nationalist–communist opposition. In the interim Yeltsin and Kozyrev responded flexibly to internal pressures and moods. There was no sudden lurch to nationalism at the beginning of 1994, after Zhirinovsky's successes, but a stage-by-stage evolution in that direction since 1992. It was in February 1993 that Yeltsin asked the United Nations to grant Russia 'special powers' as a guarantor of peace and stability in the former Soviet area.[12] In the course of that year the Foreign Ministry increasingly acted in concert with the military to forge a Moscow-dominated economic and security community in the CIS. A new assertiveness also appeared in relations with the west, as Moscow bargained harder on trade issues, and began to take independent initiatives in areas like the former Yugoslavia. The process continued, with a further hardening, warnings of a 'Cold Peace' and a reversal of the decision to go ahead with participation in the NATO Partnership for Peace at the end of 1994.

It is important not to oversimplify. As the atmosphere of internal confrontation around foreign affairs has dissipated, so policy has acquired a pragmatic, multi-faceted character, reflecting the various competing interests and institutional perspectives which it has been adjusted to satisfy. Thus Russia's more independent line *vis-à-vis* the west in 1994 was balanced by continuing efforts to join western institutions. Preparations went on for accession to the Council of Europe; Moscow agreed to sign a far-reaching Partnership and Cooperation Agreement with the European Union. In March 1995 the Russian government satisfied the requirements of the IMF sufficiently for it to release a further US$6.4 billion loan. Publicly refusing to countenance the idea of NATO expanding to the east, but excluding Russia, the Foreign Ministry nevertheless negotiated in private over the conditions under which this expansion might take place. In May 1995, Russia renewed its involvement in the Partnership for Peace. Moreover, while Moscow in practice obstructed wider OSCE participation in peacekeeping in Nagorny-Karabakh, it continued to accept the idea in principle.

In regard to the CIS, too, Russian policy reflects compromises and a degree of uncertainty. There is little willingness in most quarters to take on the economic and political costs that a thoroughgoing consolidation of the CIS would bring. Russians enjoy a markedly higher standard of living than inhabitants of Ukraine, Kazakhstan and Belarus, for example. The 'border protection' operation in which Russian troops are involved in Tajikistan provokes frequent references in the press to the disastrous intervention in Afghanistan. In most cases Moscow prefers to make use of the leverage afforded by its relative economic strength, its control of transport links and the presence of large Russian-speaking minorities in the other states. Nevertheless direct military intervention has been used, as in the case of Georgia, when Russia judged the government concerned too weak to enforce order on its own, and on the request (however constrained the choice) of that government.

Russian near-abroad policy, then, is flexible and differentiated, and still evolving. In Ukraine and the Baltic states, concern about the west's likely response has undoubtedly had a restraining influence. In some areas (the Baltic states, Tajikistan) the fate of Russian-speakers is an important issue. In some (Tajikistan, the Baltic states and the Transcaucasus) geo-strategic calculations play an important part. In the case of Ukraine and Kazakhstan especially, there are strong economic considerations.

In crude terms Moscow appears to be bartering economic assistance in return for concessions (bases, treaties) in the security sphere, but the content and the balance of power in the dialogue differs substantially from case to case. The Kazakh and Armenian governments, for example, seem to perceive a heavy security dependence on Russia, Ukraine is concerned to preserve its military/security autonomy, but has a pressing need for economic help, Belarus is seeking a closer union on virtually any terms it can get. In many Russian minds there is a sense of an inner core (Russia, Ukraine, Belarus, and all or part of Kazakhstan) which will ultimately reunite, and an outer circle in which Russia should exercise predominant influence. In 1992 and 1993 advocates of a more active near-abroad policy in Moscow kept referring to the need for Russia to adopt its own 'Monroe Doctrine', and this gives a clue to the precedent Russian policy-makers have in mind. In view of the institutional weakness of the CIS, it is more helpful to think in terms of a set of varied bilateral relationships between Moscow and the capitals of the other former Soviet states than of a new Union emerging, or of a new 'imperial' entity being created.

How to respond?

In Russian foreign policy, as in internal politics, the over-ambitious westernising goals of 1992 were modified and more defensive, pragmatic approaches were adopted. The shocks of social upheaval and post-imperial adjustment, the problems of institution-building and constructing new relations with the other former Soviet states have turned out to be more difficult to cope with than was foreseen. At the same time, policy became more broad-based and consensual, so that there is much less likelihood of fundamental change in the mid-1990s even with a new government and president.

In such circumstances it is tempting to think that there is little the west can do to influence the course of events: factors internal to Russia and the CIS will determine everything, and the wisest course is to keep on the sidelines until the process settles down. This would be to ignore, however, the special features of Russia's relationship with the West, and the recent history of the Russian elite's perceptions of that relationship, as it tries to define a new national identity and place in the world. In both regards events reached a watershed in 1995: Moscow could persist, albeit more cautiously, with its programme of opening-up and westernisation, or it could slide back into another cycle of confrontation and isolation. The possibility of an extremist turn in politics still cannot be ruled out. A good case can be made that the west missed chances to push developments inside the country in its direction in the years 1989 to 1992, when the Russians were in an unusually receptive mood. Now the opportunities to make a difference are much more limited. However it is possible at least not to make things worse.

The radical opening-up which occurred over the last decade means that there are innumerably more channels of contact between the former Soviet states and the western countries, and a large number of interests are engaged on both sides. Many agreements have been reached which offer substantial mutual benefits. Interdependence does not automatically bring peace, as New Political Thinking rhetoric implied, but it does constrain partners in their actions, and this now applies on both sides. Even in the Cold War period the west derived advantages from detente and economic ties that made conditionality easier to talk about than to enforce.

Thus Russian sensitivity, unpredictability and the denser pattern of relations help to create a series of policy dilemmas for western partners. The most pressing of these concern institutional issues, in particular the role of NATO and the EU.

Given the uncertainty surrounding the future of Russia and the CIS, and taking into account Russia's sizeable remaining military capability, it would be imprudent to dismantle the Atlantic Alliance, or to replace it by a wider body in which Moscow would have blocking powers. On the other hand, Russians still tend to see NATO, and especially plans for its enlargement, as an expression of balance-of-power, confrontational thinking. Thus, some argue that by seeking to guard against a possible future threat, we may be helping to create it.

At the ministerial session of the North Atlantic Council in December 1994 NATO member states committed themselves to a widening of the alliance, embracing probably Poland, Hungary, the Czech Republic and Slovakia in the first instance. Shortly afterwards the Russian government declared that it was suspending participation in the Partnership for Peace, and launched an invasion of the Chechen Republic, in defiance of world opinion. The prospect of NATO expansion gives a powerful argument to those parts of the Russian military and the beleaguered defence industry which are casting around for justifications for reversing the continual slide in their share of the budget. The KGB's successor, the Federal Security Service, now speaks more confidently about the dangers of western subversion.

NATO enlargement plans appear to be lending extra impetus to Russia's efforts to construct its own military bloc in the CIS area. In this sense Ukraine, for example, looks on NATO expansion with mixed feelings. Moscow will also be tempted to try to win allies in those parts of central- and south-eastern Europe not admitted to NATO in the first wave of new members. In April 1995 it was reported in Sofia that Russia had offered Bulgaria security guarantees, including the protection of its nuclear umbrella, provided it remained neutral.[13] This move is probably best understood as part of the negotiating game with NATO, but in future such manoeuvres could have serious implications. Thus by attempting to protect a region relatively free from outside threats, the west risks increasing insecurity in regions further to the east.

Since 1989 international relations among the former Warsaw Pact states have been substantially demilitarised, to the overwhelming benefit of the west and the states freed from Soviet domination. Any move which appears to reverse that tendency inevitably has the effect not just of strengthening the influence of the military and associated groups in Russian politics, but also strengthening Russian influence inside and on the periphery of the CIS and more widely. In a struggle for international influence using economic and cultural means Russia has few cards to play, with the exception of its hydrocarbon reserves. In the military sphere it still possesses a very strong hand.

In this light it makes sense to hasten slowly with NATO expansion. After all, the harmonisation programme envisaged by the Partnership for Peace would help to make a rapid expansion of the alliance more feasible, should a turn to aggressiveness in Russian policy make it necessary in future. In the meanwhile, at a moment when Russia is still groping for a role in international affairs, we should avoid handing it a script in which it plays the part of chief enemy in Europe and gendarme in Asia. As NATO expansion moves ahead, it is essential to combine it with energetic countervailing moves to counter the impression that Russia and the CIS are being shut out of Europe, and that new barriers are being erected. The most reassuring route would be to build a close partnership between Russia and NATO.

During the negotiations over participation in the Partnership for Peace in 1994, Russian was offered an enhanced relationship with NATO, but there has been little enthusiasm on the western side, where it is feared that Moscow aspires to exercise a veto over NATO decisions. NATO needs to find the right point between two extreme solutions. One would be to allow Russia into the discussions only once a joint position had been settled between all NATO members (which would leave little flexibility to take its views into account). The other would be to allow access at an early stage in policy deliberations, something which would clearly be unacceptable to most NATO states. A study by RAND analysts published in early 1995 suggested instituting regular meetings between Russia and key NATO bodies such as the North Atlantic Council, the Defence Policy Committee and the Nuclear Planning Group, and replicating the model of the Bosnia Contact Group (which brings together the United States, the United Kingdom, Germany, France and Russia) in other fields.[14] Formalising such innovations in NATO's relations with Russia, for example in a new treaty, would increase the domestic political pay-off there.[15]

Hopes that CSCE (now OSCE) would serve as the basis for a new over-arching European security system have largely faded in Moscow as elsewhere in Europe. Nevertheless Russia continues to propose that the institution should be strengthened, in particular by providing it with a 12-strong 'security council', and that it should be given powers of control over the activities of NATO, the WEU, the CIS and the EU (in their peacekeeping activities). While subordination of NATO (and the CIS for that matter) to the OSCE is clearly out of the question, reinforcing its importance in general is attractive because it would help to counteract the tendency to establish two watertight military blocs in Eurasia, and could be used as a lever against the application of military means inside the CIS. If the smaller OSCE member states could be persuaded to accept rotating membership of an inner council alongside seven or eight permanent members, such an innovation could enhance communications and build trust, as well as regularising Russian involvement in European security debates. It would be particularly reassuring for a state like Ukraine if it could participate as one of the core members.[16]

The European Union is in many ways a more suitable framework than NATO for 'spreading security to the east'. Because of its economic dimension it has tended to appear to Russians more an engine of integration than as an engine of confrontation. Second, it can offer varying degrees of association: states are not simply 'in' or 'out'. Potential new dividing lines are blurred and the likelihood of new military blocs crystallising is lessened.[17] Russia is not a candidate for full membership of the EU, but from the point of view of offering reassurance to westernisers in Moscow and fostering a sense of the country's European destiny, it makes sense to emphasise the possibility of a close relationship. In the shorter term introducing freer access to west European markets would send the right signals to managerial groups which form an important 'swing constituency' in Russian elite politics.

A variety of benefits can come from these institutional links. An important one is wider face-to-face contact, which helps to counteract the effects of seventy years of near-isolation in the former Soviet Union, and attacks directly the societal roots of a potential slide back into confrontation. This is most obvious in the case of the military trust-building processes which have gone on in the framework of NACC.[18] It strengthens the case for extending technical assistance and educational cooperation programmes, and in general for applying some of the techniques used in Western Europe in recent decades to build a 'citizens' Europe', with a proper recognition of the enormous size of the task.[19] The accession process to the Council of Europe which got under way in 1994 involves a sizeable amount of contact at all levels.

The task which faces us can be seen as one of 'socialising' a country which has got used to breaking the rules into playing a more normal part in international society. This means laying down clear standards of behaviour, offering appropriate incentives and deterrents, avoiding the twin dangers of permissiveness and reducing the level of contact so far that the scope for leverage is lost. The decision to suspend the Council of Europe accession process after Russia's invasion of Chechnya at the end of 1994 was undoubtedly correct. The need to enforce standards and to maintain credibility outweighed the prospective benefit of Russian entry to the Council of Europe. To join a law-respecting society, Russia must first respect the law.

But is also important not to make impossible demands. The arguments about what type of economic 'therapy' is appropriate to the ex-Soviet societies are too extensive and too familiar for them to be repeated here. It is worth commenting, however, on a more specifically security-related economic issue which touches on the interests of an influential part of the Russian elite. The partial demilitarisation of the former Soviet economies has had a devastating effect, especially in those areas of Russia and Ukraine where the defence industries were concentrated. An unusually high proportion of Russia's high-technology capacity has developed in association with military research and development, and the defence sector is widely regarded as the core of the country's industrial potential. It is a vital interest of the west's to prevent uncontrolled spreading of military technology, but too restrictive an approach to Russian arms exports, or one which appears to involve double standards, risks a boomerang effect, and a discrediting of the idea of an international regime altogether.

The case of CFE flank equipment limits (discussed in greater detail in Chapter 14) involves a similar calculus. CFE limits set in 1990 were based on the assumption of a continuing Soviet Union, and certainly did not take into account the kind of military upheavals which are taking place in the Caucausus. In Russian eyes the limitations on southern flank forces appear arbitrary and, what is more, hamper the army in its most fundamental task, to preserve Russia's territorial integrity. This provokes Moscow to bend or break the rules and puts the arms control process at risk.

Intra-CIS relations present one of the biggest challenges to western judgement. First of all, the experience of the former Yugoslavia makes it incredible that the United States or any of the west European powers would allow themselves to become involved in any substantial way in military operations in the former Soviet region. Thus enforcing peace (which unfortunately appears to be necessary in many places) becomes a task for CIS (*de facto* Russian) forces. At the same time, it is important for western governments to hold Moscow both to its commitments to international law (especially respect for the sovereignty of the newly independent former Soviet republics) and to specific commitments undertaken by Russia since the breaking up of the USSR.

On the one hand Russia can easily exert invisible pressure by supplying arms, employing covert measures, manipulating Russian-speaking communities abroad, and threatening to cut off economic subsidies of one kind or another. On the other there are genuine centripetal forces at work. These may be connected with the immediate economic interests of industrial communities cut off from sources of supply and markets in the former Soviet heartland, or with the cultural affinity which many of the ruling and managing elites in the other states feel with Russia. This combination makes it difficult to judge how voluntary are moves to rapprochement between Russia and its former Soviet partners.

Western governments should not only tolerate but encourage closer economic ties inside the CIS because they are a prerequisite for recovery of living standards to the minimum level needed to sustain civilised democratic practices. Moreover, there is no necessary zero-sum relationship between intra-eastern and east–western economic integration, and it is important to make that clear.

The complexities of Russian domestic politics also create difficult choices. Some argue that western governments should turn a blind eye to creeping authoritarianism for the sake of having irreversible socio-economic changes pushed through by a leadership which declares its underlying attachment to democratic values. But it is more important to keep the way open for a healthy alternation in power. The actions of competing political groups are likely to be more moderate when they come to office than their rhetoric sounds in opposition.

Conclusion

The idealistic image of future relations with the west which was used by Soviet reformers to justify calling an end to the Cold War and dismantling the eastern bloc raised impossible expectations. It was the Cold War itself, paradoxically enough, which was largely responsible for the cooperativeness and solidarity in intra-western relations which the New Political Thinkers had observed and wished to extend to a pan-European scale. But with more imaginative and more courageous leadership more could have been done to sustain the hopeful atmosphere of the first years of the decade.

Habits of thought and organisational interest create a powerful inertia around existing institutional arrangements which is difficult to overcome. NATO and the European Union have adapted reasonably well to central European demands, but disappointingly little progress has been made towards creating a clear framework into which the former Soviet states can fit. Lack of consultation and what are perceived as attempts to make unilateral geo-strategic gains by the western powers have generated resentment in Moscow. In the sphere of economic policy advice, western governments did not pay enough attention to special local political and social circumstances in the CIS, and states have been encouraged to apply 'universal' models of reform, with very disappointing results.[20] In the sphere of trade and economic cooperation little has been done to demonstrate that abandoning Soviet-style autarky brings a positive pay-off.

There has thus been a growing tendency in Moscow to think more in terms of balance of power and less in terms of integration. From a certain point of view this may be welcome, as a sign that Russian perceptions are maturing and settling down. However, Russia can not hover indefinitely on the periphery of Europe, as some kind of neutral force. In the end either it must be integrated into European structures, or Europe must prepare to defend itself against it. Neither of these options is inexpensive, but the first is undoubtedly preferable.

11 DEVELOPING THE CONFLICT PREVENTION AGENDA[1]

Andrew Cottey

The concept of conflict prevention—or preventive diplomacy—has come to the fore of international attention since the end of the Cold War. The reasons for this are hardly surprising. Successful conflict prevention could save many lives and avert much human suffering. It could also help the international community to avoid difficult (indeed, sometimes impossible) choices in responding to conflicts. As the wars in former Yugoslavia starkly illustrated, once violence breaks out the international community is often faced with a choice between a range of policies of limited efficacy—mediation, diplomacy, political pressure, economic sanctions, humanitarian aid—and costly and risky military intervention. Conflict prevention could be far more cost-effective than responding once violence has broken out. Fact-finding, mediation, monitoring of human and minority rights, early use of political and economic sanctions and deploying small military forces in trip-wire roles are relatively inexpensive when compared with the costs of large-scale humanitarian aid, relocating refugees, peacekeeping operations or military intervention.

The British government endorsed the concept of conflict prevention in September 1994, when Prime Minister John Major declared:

> For years, our energies have been consumed in trying to limit trouble after it has started. But it would be far, far better—and indeed, far less costly—to pre-empt it....an entirely new effort at preventive diplomacy is long overdue....Britain wants to develop new mechanisms to head off conflicts before they become unstoppable.[2]

Given its position as a permanent member of the United Nations (UN) Security Council and a leading member of the European Union (EU), the North Atlantic Treaty Organisation (NATO) and the

Organisation for (formerly Conference on) Security and Cooperation in Europe (OSCE) and its substantial diplomatic and military experience, Britain has the potential to contribute significantly to the development of more effective means of averting conflicts. Nevertheless, as UN Secretary-General Boutros Boutros-Ghali has observed:

> For all its historical and currently near-universal acceptance as a vital field of action, preventive diplomacy remains an elusive topic. The concept needs to be expanded. Steps for elaborating and employing preventive diplomacy in a sustained and effective manner remain to be taken.[3]

This chapter explores the policy challenges raised by conflict prevention, focusing on the role of the UN, but highlighting areas where Britain can make a particular contribution.

Early warning, agenda-setting and policy-formulation

Effective action to prevent conflicts depends on knowing where and when conflicts are likely to occur (early warning) and on ensuring that potential flashpoints are put on the agenda of policy-makers (agenda-setting) and that appropriate policies are developed (policy-formulation). International institutional mechanisms for early warning, agenda-setting and policy-formulation, however, remain relatively underdeveloped.

Following a 1982 initiative by Secretary-General Javier Pérez de Cuellar the UN set up an Office for Research and Collection of Information (ORCI) in 1987.[4] Due to a lack of political will amongst the member states and an unclear mandate, however, ORCI did not develop a significant role. Boutros-Ghali dismantled ORCI early in 1992 and its work was subsumed into the newly founded Department of Political Affairs, whose role is to provide early warning and analysis to support preventive diplomacy and conflict resolution.[5]

The work of the UN Secretariat and the Department of Political Affairs, however, remains overly focused on conflict management. High-level political attention and individual desk officers' time is, not surprisingly, focused on dealing with existing conflicts. The problem is compounded by the division of labour between the different components of the UN Secretariat (particularly the Departments of Political Affairs, Humanitarian Affairs and Peacekeeping Operations) and the various UN specialised agencies which means that none has a comprehensive overview of relevant developments. One possibility would be the creation of an Office of Preventive Diplomacy within the UN (perhaps located within the Office of the Secretary General)—a high-level policy planning unit mandated to monitor and analyse potential conflicts and develop policy options. Such a relatively small, high-level body could play an important role in providing early warning of conflicts, putting potential conflicts on the political agenda and formulating policy options.

In *An Agenda for Peace,* Boutros-Ghali noted that, while the UN had an increasingly diverse range of information sources, there remained a need 'to strengthen arrangements in such a manner that information from these sources can be synthesised with political indicators to assess whether a threat to peace exists and to analyse what action might be taken by the United Nations to alleviate it'.[6] Boutros-Ghali's call generated technical studies of the requirements of a UN early warning system as the Secretary-General consulted with regional organisations and member states on the issue.[7] Within the UN, an inter-agency group is studying the issue of data management and early warning. If conflict prevention is to be given greater attention, however, early warning will have to be integrated with analysis and policy formulation—a role which an Office of Preventive Diplomacy could play.

The central problem is not the collection of data but the ability to collate and analyse it.[8] The primary role of a UN conflict early warning system should be to synthesise existing information from a variety of sources in order to develop a clear picture of potential conflicts and provide policy recommendations. While this already occurs to some extent within the Department of Political Affairs,[9] its ability to provide conflict early warning is limited by the focus of its attention on ongoing conflicts and its lack of an inter-departmental overview. An effective conflict early warning system requires that information from a host of relevant sources—other UN agencies, national governments (including Foreign Ministries and intelligence agencies), regional organisations, non-governmental organisations—is collected and analysed in an on-going, systematic way. Equally important, whichever component of UN Secretariat is given the task of providing early warning of conflicts also needs to have a clear mandate to produce critical analysis and policy options and recommendations in order to inform the deliberations of the Secretary-General, the Security Council, the General Assembly and regional organisations.

Whilst early warning is important, it is of little use unless it leads to action. An effective system of conflict prevention must put issues on the international agenda and formulate policy. The Secretary-General has the power, under the UN Charter, to draw the Security Council's attention to any potential threat to international peace and security.[10] Although the General Assembly and the Security Council have endorsed the idea that the Secretariat should provide regular early warning and policy recommendations, this has not yet happened on a systematic basis. The Bush administration shared intelligence reports with EU governments in 1990 that suggested civil war in Yugoslavia was imminent, but the EU was ill-prepared to act when conflict erupted in 1991.[11] Agenda-setting and policy-formation could also be strengthened by the establishment of an independent monitoring group (composed of a small group of states selected by the General Assembly) designed to act as a 'trigger mechanism' ensuring 'the automatic debate of a crisis'.[12]

The Security Council could also enhance its capability to deal with its expanding agenda, by increasing the number of permanent members. The creation of regional sub-committees of the Security Council (linked to relevant regional organisations) to deal with potential conflicts which have not yet escalated to a level where they require high-level political attention could provide a further focus for developing preventive policies.

European mechanisms for early warning, agenda-setting and policy formulation also need to be further developed. At the November 1990 Paris summit, the then-CSCE established a Conflict Prevention Centre (CPC). The CPC, however, lacked adequate resources or a clear mandate and mechanisms for providing early warning or putting potential conflicts on the CSCE's agenda. As a result, the Committee of Senior Officials (CSO), rather than the CPC, became the primary focus for political discussions within the CSCE. In December 1993, the CSCE established a Permanent Committee, creating a standing mechanism for the discussion of potential and ongoing conflicts, which was upgraded to the Permanent Council at the December 1994 Budapest summit of the CSCE (now renamed the OSCE). These OSCE structures are supported by a number of mechanisms allowing various actions to be taken in response to an actual or potential conflict.[13] These structures could, however, be further strengthened with regular and ongoing assessments of potential conflicts and policy options for it to discuss. If the OSCE is to develop a more effective role it needs to move further away from the principle of unanimous decision-making. Two routes may be possible. One would be to establish a Security Council somewhat akin to the UN Security Council, with powers to authorise diplomatic and economic sanctions and possible military enforcement. An alternative would be to establish some form of qualified majority voting (QMV). While progress in these areas will not be easy, it would be necessary if the OSCE is to develop a more effective role.

The EU also has the potential to play an important role in preventive diplomacy. The fact that many states in eastern Europe, the former Soviet Union, the Mediterranean and Africa are pursuing membership of (or expanded political and economic ties with) the EU gives EU governments the potential to influence significantly those states' domestic and foreign policies. Through its stability conference, the EU is already using political conditionality for economic ties and membership to help to moderate disputes between Hungary and its neighbours and the Baltic states and Russia. The ability of the EU to formulate consistent and effective policies for conflict prevention could be much further strengthened, however. In particular, regular meetings of EU Foreign Ministers need high quality early warning, political analysis and policy options to underpin their decisions. The 1996 intergovernmental conference (IGC) on the future development of the EU could therefore mandate the Commission (in conjunction with national Foreign Ministries) to establish structures for the provision of such early warning, analysis and policy options. EU governments also need to think through in a more strategic manner how to use growing political and economic influence to support preventive diplomacy. Again, the 1996 IGC could mandate the Commission to undertake a detailed study of the policy instruments available to the EU for conflict prevention and ways in which those instruments could be more effectively used and further developed.

Developing policies for conflict prevention

While early warning, agenda-setting and policy-formulation are important, they are only of use if they lead to the development and implementation of effective policies. The policy instruments available to the international community for conflict prevention could be strengthened and further developed in a number of ways.

Fact-finding, mediation and diplomatic missions

Conflicts are easiest to resolve at an early stage in their evolution, before opposing positions have polarised. The most appropriate first line of action is a low profile fact-finding, good offices or mediation mission. Within the UN, such missions may be undertaken at the initiative of the Secretary-General, in the role of neutral mediator (although with an implied threat of further action if the conflict is not resolved).[14] Since the end of the Cold War, the number of such missions has expanded significantly, with over 100 in 1992–93.[15] In September 1993 the United Kingdom and France offered to provide personnel and communications equipment for such missions. In his January 1995 supplement to *An Agenda for Peace*, however, Boutros-Ghali highlighted the continuing 'difficulty of finding senior persons who have the diplomatic skills and who are willing to serve for a while as special representative or special envoy of the Secretary-General'. He also highlighted the need to develop 'accepted and well-tried procedures' for the establishment of full-time field missions to support preventive diplomacy and to provide the necessary financial support for preventive diplomacy missions.[16] One way to address this problem could be to build on the 1993 Anglo-French initiative by establishing a stand-by system (similar to the system currently being developed for peacekeeping forces), with states agreeing in principle to make personnel and equipment available for UN missions at short notice and the establishment of a dedicated budget. Given that such missions are far less costly and risky than peacekeeping missions and are less likely to involve an indefinite commitment, such a preventive diplomacy stand-by system is perhaps more likely to be successful than the current peacekeeping stand-by system.

If a relatively low-profile mission fails to resolve a conflict, the next step is a more high-profile mission authorised by the Security Council or the General Assembly. The Secretary-General, Security Council or General Assembly may continue to play the role of a neutral mediator, but acting in a more high-profile manner.

Alternatively, they may recommend or support a particular resolution to the dispute. In situations where neutral mediation efforts have already failed, however, successful mediation is likely to depend on the ability of the international community to bring pressure to bear on one or more of the parties to the dispute. (The differing ways in which this may be done are examined below.)

Human and minority rights

One of the defining characteristics of many of the conflicts of the post-Cold War era is an ethnic or inter-communal dimension. A central element of strategies to prevent such conflicts should be educational systems which eliminate provocative stereotypes and generate tolerance for diversity of race and religion and measures to protect human and minority rights. Action at the earliest possible stage is central: 'a comparison of cases in which tensions were either addressed or ignored in early negotiations underscores the importance of providing for the concerns of various ethnic groups, even in situations in which there is no apparent tension'.[17]

Since the Second World War, the international community has made significant steps in defining individual human rights and, to a lesser extent, in developing forums for monitoring those rights (such as the UN's Economic and Social Council and Commission on Human Rights). The ability to respond to violations of human rights, however, remains relatively limited.[18] More effective monitoring of human rights and action against states violating them could help to prevent conflicts. The already growing use of UN Special Rapporteurs (appointed by the Commission on Human Rights) and Special Representatives (appointed by the Secretary-General) to monitor human rights and report violations to the General Assembly and the Security Council could be expanded on a systematic basis. The Commission on Human Rights might also adopt an emergency procedure involving a permanent register of experts who could be dispatched as 'white helmets' to investigate claims of gross human rights violations and report violations to the Security Council.[19] Linkage of these mechanisms to the Security Council would increase pressure on states violating human or minority rights and create the possibility of further action if violations continued.

Central to many current conflicts are issues of ethnic group rights and the balance of power between different ethnic groups within states. The development of mechanisms for protecting minority rights could contribute significantly to conflict prevention. Mechanisms might include forms of regional autonomy, electoral systems designed to balance the power of different groups, guarantees of specific rights, joint exercise of power by differing ethnic groups, cultural autonomy, proportional representation for minorities or a minority veto.[20] The provision of expertise, mediation and political pressure by the international community could contribute to the development of such mechanisms. In Europe, the Council of Europe and the OSCE's High Commissioner on National Minorities already play a significant role in this area. In the UN context, the Sub-Commission on Prevention of Discrimination and Protection of Minorities (a sub-body of the Commission on Human Rights) could undertake a long-term commitment to explore differing means of protecting minorities and provide suggestions to states, the Secretariat and the Security Council.

Ethnic conflicts also raise the issue of secession. Given the conflict-generating potential of secessionist claims, most international efforts should be directed towards protecting human and minority rights within existing states, rather than supporting secession. Nevertheless, there is a need to consider the circumstances in which support for the right to self-determination may be appropriate. The Security Council could consider setting up an international commission to make recommendations in relation to secessionist claims and possible guarantees of minority rights. Criteria for secession could include: clear evidence either that peace cannot be maintained within an existing state or that minority rights are being seriously violated; a democratically expressed wish to secede; commitments to respect human and minority rights in any new state; viable state borders for new states; and reasonable prospects of political and economic stability for the new state.[21] A state's refusal to accept the jurisdiction of such a commission could trigger further action.

Confidence-building measures and arms control

Regional confidence-building measures and arms control agreements also have a role to play in conflict prevention by improving political relations and reducing misperceptions and risks of accidental war. The most developed regimes in this field are in Europe, within the

OSCE context, although these have proved ineffective in influencing the behaviour of the newly independent republics of former Yugoslavia. In the Middle East, disengagement agreements between Israel and Egypt and Israel and Syria include de-militarised zones patrolled by outside forces, force-limitation zones constraining the deployment of troops and offensive equipment, direct and third-party overflights over the zones and international inspections.[22] India and Pakistan negotiated similar, though less extensive, measures. Other possible measures include bilateral exchanges of information on military doctrines and forces and military-to-military contacts. The UN, regional organisations and individual states can support such regimes through the provision of expertise, support for implementation and verification and diplomatic pressure to develop such regimes.

To date, these types of measures apply only to inter-state conflicts. Given the predominance of intra-state tension, such measures need to be adapted to fit potential intra-state conflicts. In situations where the international community is already intervening within a state in a mediation role, for example, it might be possible also to pursue the withdrawal of armed forces, both government and irregular, from particularly sensitive regions. The UN or regional bodies could provide neutral forces for the implementation and verification of de-militarised zones within states.[23] Obviously, sensitive issues of state sovereignty may be raised, but this does not mean that the possibility should be rejected.

As noted in Chapter 6, constraints on the international arms trade could also contribute to conflict prevention. While simplistic arguments that arms races cause conflicts are inaccurate, the ongoing supply of arms to regions of potential conflict certainly exacerbates tensions and may be one factor triggering violence. The EU and the OSCE have both agreed criteria governing conventional arms transfers.[24] Relatively little progress has been made, however, in putting such criteria into practice. The next step ought to be agreement on what such criteria mean in practice and the development of lists of countries to which particular exports would be embargoed or controlled.[25] At the global level, the UN Register of Conventional Arms (established in 1991) has introduced a new degree of transparency into the international arms trade. However, the register has no mechanism to ensure that states submit returns or that information is accurate, and does not yet include small arms. The development of a mechanism to put much greater pressure on states to participate and the inclusion of small arms would enhance the register's utility as a tool for monitoring the arms trade and hence constraining destabilising transfers. Britain also bears a heavy responsibility to curb its overly aggressive export of arms.

Political, economic and military pressure and sanctions

In many potential conflict situations, simply providing neutral mediation, expertise or policy recommendations is not enough. Mediation efforts must be supported by a combination of political, economic and military muscle that promises (or threatens) an effective mix of pain and gain.[26] In this context, Britain and other major western powers, given their political, economic and military power, have an important role to play. Political and economic pressure at an early stage in a conflict's development may often be the most effective way to prevent escalation to violence. Once conflicts have polarised and positions hardened, even highly coercive forms of pressure may have relatively little chance of altering positions.

Britain should use what influence it retains in the international community to develop and utilise a more sophisticated range of ways of bringing pressure to bear on parties to potential conflicts. Political and diplomatic ties, for example, may be made conditional on states' respect for human and minority rights and pursuit of foreign and defence policies which do not threaten or provoke conflict with their neighbours. Similar conditions could be attached to long-term economic aid channelled through such bodies as the IMF and the World Bank and from national governments. Although this might require changes to the current mandates of these bodies, such a step could make a significant contribution to conflict prevention. In more serious cases, the Security Council could consider selective economic sanctions against states deemed to being provoking or exacerbating potential conflicts.

There is also a need to explore how economic pressure may be graduated in relatively sophisticated ways to put pressure on parties and provide positive incentives for the resolution of conflicts. The US Congress's threat to reduce economic aid to El Salvador, for example, put pressure on that country's military to compromise in the UN-sponsored peace negotiations, contributing to a successful outcome.[27] More effective and sophisticated targeting of economic sanctions could increase their utility as a tool for conflict prevention. Early sanctions on transfers of particular goods, technologies and forms of finance might, for example, be effective in putting pressure on parties to potential conflicts (especially central governments and armed forces). Since many countries in which conflicts are likely are recipients of military aid from major powers, the linkage of such aid to policies bearing on potential conflicts could also have a significant impact.

Finally, the threat of military intervention remains an option for conflict prevention. In Haiti, it may be argued, only the threat of US military intervention forced the Haitian military leadership to step down in 1994 and averted further bloodshed. Obviously, threats of military intervention raise highly contentious issues and are likely to be a last resort. Nevertheless, as Hugh Beach argues in Chapter 12, there may be circumstances in which an early threat of military action may be able to prevent a particular conflict.

Preventive deployment

In *An Agenda for Peace*, Boutros-Ghali suggested that preventive deployment might occur at times of crisis within a country (at the request of the government or all parties concerned, or with their support); at the request of two states involved in an inter-state dispute; or, when one state feels threatened and the Security Council determines that a deployment along its border is appropriate.[28] In December 1992 the Security Council authorised for the first time the preventive deployment of a UN peacekeeping force into a country where conflict had not previously broken out—to Macedonia, on its borders with Albania and Serbia.[29] Although Macedonia remains a potential area of conflict, the UN force appears to have helped to reduce tensions.

Both the concept and the practice of preventive deployment, however, remain relatively underdeveloped.[30] In particular, there is a need for more consideration of the circumstances in which UN forces may be deployed preventively, the mandate under which they should operate and resulting force requirements. Expanding on Boutros-Ghali's three scenarios, a number of options may be possible. In inter-state disputes, traditional peacekeeping forces might be deployed, playing a role in promoting confidence and defusing low-level incidents, but with their presence dependent on the consent of the parties.

However, if the problem is the risk or fear of one state deliberately attacking another, it may be more appropriate for the Security Council to authorise the deployment of a force mandated to use force to respond to an act of cross-border aggression.

Preventive deployment within states raises different problems. In most cases, the central requirements are likely to be the consent of the government concerned and the acquiescence of other parties to a potential conflict. In this context, a number of possibilities may exist: traditional peacekeeping operations designed to reduce tensions, build trust and promote negotiations; international civilian forces to monitor human and minority rights developments; and elements of enforcement (for example, to demobilise military forces or enforce demilitarised zones). The fear of being dragged into civil wars is likely to limit how far the Security Council moves in the direction of enforcement. Where state authority has collapsed or is severely weakened, one possibility may be the creation of a UN trust territory supported by the deployment of UN forces whose mandate includes elements of peace enforcement[31]—as some observers suggested should have occurred in Bosnia prior to the outbreak of the conflict there in 1992.

The concept of preventive deployment is at an early stage in its development. Given the reluctance of UN member states to provide forces for existing peacekeeping operations, the prospects for its more widespread use may be limited. However, since a central feature of preventive deployments may often be a tripwire function, the forces required in such cases ought to be relatively small and lightly armed. Even if significantly larger forces may be required, successful preventive deployment would justify the costs and risks involved, averting the need to intervene with much larger peacekeeping or peace-enforcement forces and mount expensive humanitarian relief operations once violence has broken out.

Expanding the conflict prevention agenda?

While the types of policies suggested above could make an important contribution to the prevention of many conflicts, they address the proximate, rather than the underlying, causes of conflicts. An effective conflict prevention agenda must also address the more fundamental causes of conflicts. As Thomas Weiss has argued:

The root causes of many conflicts—poverty, the unjust distribution

> of available resources, and the legacy of colonial boundaries in many multi-ethnic societies—should be addressed before they explode. Effective prevention requires investment in economic and social development in poor countries as well as reforms to distribute the benefits of economic growth more equitably. It also would include reform of the global finance and trading systems.[32]

Clearly, the development of policies to address these issues will not be easy or achieved overnight. There is, however, growing recognition that the task must be addressed. The March 1995 UN Social Development Summit in Copenhagen established the principle that IMF and World Bank policies must be judged not only by their ability to produce economic growth, but also by their ability to support the provision of basic (economic) rights for the world's poor. The summit also enshrined the principle that sustainable development must include civil, political, social, economic and cultural rights.[33] While progress in putting such policies into practice will not be easy, the establishment of the principle is important.

In this context, Britain should consider how the structures of the IMF and the World Bank can be reformed to address this longer-term conflict prevention agenda. The specific criteria by which the IMF and the World Bank give aid should be expanded to include not simply economic growth, but also the alleviation of poverty, the provision of basic social needs and respect for human rights and democracy. More broadly, the wealthy western states must recognise that continued impoverishment underpins civil and regional conflicts, the growth of religious and nationalist fundamentalism and the proliferation of conventional and non-conventional weapons in much of the Third World. In the longer term, if an intensifying North—South conflict and a growing 'clash of civilisations' are to be avoided, the underlying causes of these problems must be addressed.

Conclusion

This chapter has explored the challenges posed by conflict prevention in the post-Cold War world. It suggests a number of forward-looking but pragmatic steps to enhance the international community's ability to prevent violent conflicts and address the underlying causes of conflicts.

Britain is well placed to play an important role in developing and implementing this conflict prevention agenda. As a permanent member of the UN Security Council Britain could press for the development of more effective UN early warning and conflict prevention structures. Within the EU and the OSCE, Britain should urge new structures to develop and implement preventive policies. Britain could also use its significant diplomatic experience to underpin international mediation and preventive diplomacy efforts, through the provision of personnel, expertise and training. Britain has already taken the lead in developing regional confidence-building measures and arms control regimes, but could do much more to provide expertise and support in terms of monitoring human and minority rights. Rather than use British embassy staff to promote arms sales abroad, for example, they could be better employed to promote human rights.

The high quality and experience of the British armed forces makes them well-suited to participate in possible preventive deployment operations. As a member of the Group of Seven (G7), Britain can also play an important role in shifting attention towards the need to address the underlying economic and social causes of conflicts. The further development of the concept and practice of conflict prevention should be a central task for British foreign and security policy in the 1990s and beyond.

12 NEW COALITIONS FOR PEACEMAKING

Hugh Beach

> We have entered a time of global transition marked by uniquely contradictory trends. Regional and continental associations of states are evolving ways to deepen co-operation and ease some of the contentious characteristics of sovereign and nationalistic rivalries. National boundaries are blurred by advanced communications and global commerce, and by the decisions of states to yield some sovereign prerogatives to larger, common political associations. At the same time however fierce new assertions of nationalism and sovereignty spring up, and the cohesion of states is threatened by brutal ethnic, religious, social, cultural or linguistic strife. Social peace is challenged on the one hand by new assertions of discrimination and exclusion and, on the other, by acts of terrorism seeking to undermine evolution and change through democratic means.
>
> (Boutros Boutros-Ghali, *An Agenda for Peace*, United Nations, New York, 1992)

This chapter asks what, if anything, can Britain and the western allies do about war in its contemporary form of inter-communal conflict now becoming a universal security dilemma? Some analysts argue that there are no magic potions to remove the taste, once acquired, of ethnic disputes and too few cases of successful ethnic conflict management on which to build optimism, because when ethnic conflicts break out they are by nature nasty, brutish and *long*.[1] Yet the international community *must* do what it can to deter ethnically based groups from establishing dictatorships of nationality and undermining the prospects for democracy in multi-ethnic territories. Any force which seeks to intervene, whether foreign or multinational, will have objectives defined in terms of peace, justice or humanitarian relief. There may also be an element of national interest involved. For these goals outsiders will be prepared to accept some losses and to pay some costs. But if they find themselves in conflict with an ethnic insurgent group the struggle will almost always be unequal because the adversary will believe that he is fighting for his life. He will be prepared to pay almost any price and to accept enormous losses. Having little stake in the status quo, the insurgent ethnic group may also be willing to break any rule, convention or undertaking if this will further the cause. For guerrilla movements survival can provide a justification for almost anything. How then shall the international community seek to cope?

As the previous chapter by Andrew Cottey argues, the best remedy by far is prevention. Measures to promote economic and social development are fundamental. When things start to go wrong much can be achieved by early warning, preventive deployment of monitors and observers, conciliation and mediation. These measures have received much attention of late and figure largely in the UN Secretary General's *An Agenda for Peace*. They are *always* preferable to the use of military force. But try as the international community will, only too often matters get out of control and violence breaks out. (On 15 December 1994 *Jane's Sentinel* estimated that there were 26 such conflicts under way). The questions then to be addressed, are: under what circumstances should intervention be attempted or indeed withheld, and according to what principles should it be conducted? It is not always clearly recognised that unilateral and multilateral intervention tend to proceed according to quite different rules.

On the one hand there is the method derived from United-Nations style *Peacekeeping*, as it has developed during the past 40 years. Although there is no reference to it in the Charter, the principles of UN peacekeeping are now widely understood to be:

- operations established *by* the United Nations
- with the *consent* of the parties concerned
- to help control and resolve conflicts between them
- under United Nations command and control
- at the expense collectively of the member states, and with military and other personnel and equipment provided voluntarily by them
- acting *impartially* between the parties
- using *force to the minimum extent* necessary[2]

These are, in essence, the ingredients of classical or first generation peacekeeping: methods which scored some notable successes in such places as Central America, Eritrea, Cambodia, Namibia and Mozambique.

But since the end of the Cold War, the UN has undertaken, or subsequently blessed, a number of operations in which important features of classical peacekeeping have been abridged or totally over-ridden. Under the rubric of 'humanitarian intervention' operations have taken place in several countries where the ingredient of consent has sometimes been lacking and where impartiality and minimum force have been compromised in many instances. The theory has arisen that where a state is inflicting upon its own people gross, flagrant and continuing infringements of their human rights the international community has a right—some would even argue an obligation—to try to restrain it. The UN Charter, in Article 2.7, says bluntly that nothing contained in it shall authorise the UN to intervene in matters which are essentially within the domestic jurisdiction of any state. But it half-contradicts itself by saying that this principle shall not prejudice the application of enforcement measures under Chapter VII of the Charter. That Chapter relates not only to acts of aggression, but also to threats to the peace and breaches of the peace. Until very recently it seemed as though this criterion required that a threat to *international* peace and security be determined. Thus UN Security Council Resolution (UNSCR) 688 of 5 April 1991 described Iraqi repression of the Kurds and Shias as a threat to *international* peace and security. Though this resolution said nothing about enforcement action it was on the strength of it that France, followed by the US, Britain and a number of other countries, did take action with ground and air forces to compel the Iraqis to desist.

On 19 November 1992, UNSCR 788 determined that the situation in Liberia had become a threat to *international* peace and security, thus underwriting *post facto* the intervention two years earlier by the Economic Cooperation of West African States (ECOWAS). But in both these instances the supposed threat to international security was largely a façade. On 3 December 1992, the Security Council (in SCR 794) broke new ground by deciding to intervene in Somalia for strictly humanitarian purposes. There was not even a pretence of consent by the government of Somalia: effectively such a government did not exist. There was negligible spill-over to other countries, for example, in the form of refugees. The plight of the Somali people was the sole rationale for invoking Chapter VII of the Charter: authorising the use of 'all necessary means' to establish a 'secure environment for humanitarian relief'. A *de facto* Right of Humanitarian Intervention had thus begun to emerge as a result of these Security Council Resolutions. The task of the UN forces in Bosnia has not been described in terms of peacekeeping at all. Its task is to protect the delivery of humanitarian goods and services and to protect civilians in declared 'safe havens'.[3] The deployment of the UN Protection Forces (UNPROFOR) was carried out, initially at least, with the consent of the host states (Croatia, Bosnia, Macedonia) but its mandate was subsequently extended to include, for example, deterrence of attacks on safe havens and the occupation of key points on the ground to that end—with a clear flavour of Chapter VII enforcement.

If, in these types of operation, the principles of consent, impartiality and minimum force have sometimes been applied *in diminuendo* this has been balanced by three other principles, derived directly from the Christian 'Just War' tradition, which have come to figure more and more in the rhetoric and, to some extent, in practice: proportionality, discrimination and the concept of 'last resort'.

Proportionality
The overall destruction expected from the use of force must be outweighed by the good to be achieved.

This principle figured much in the utterances of General Michael Rose during 1994 for example. It is relevant to the US preference in Bosnia for allowing the Muslims to import arms and for air attacks on targets in Serbia of strategic importance. If foreseeable consequences included bringing existing humanitarian operations to a halt, enforcing the withdrawal of UN forces and perhaps prolonging the bloodshed this would be a bad bargain. Of course such an argument means weighing in the balance things which, even in theory, are incommensurable. How many Dutch lives was it worth to protect Srebrenica? Can one put a price on a principle? Yes, one does it every day so there is no dodging, and certainly there are no easy answers.

Discrimination or non-combatant immunity
Civilians may not be the object of direct attack, and military personnel must take due care to avoid and minimise indirect harm to civilians.

Considerable efforts were made, by the Allies in the Gulf War for example, to pay much more than lip-service to this principle. In a partisan style of conflict this is very difficult to do. In this kind of war the contestants totally and systematically ignore this principle and almost all other humanitarian laws of war. But that does not absolve the international community; quite the reverse.

Last resort
Force may be used only after all peaceful alternatives have been seriously tried and exhausted.

In other words if measures short of armed force would suffice then armed force should not be used. Articles 33 to 42 of the UN Charter describe a wide spectrum of measures available to the international community: starting with enquiry, mediation, conciliation and so forth, via diplomatic and economic measures up to demonstrations, blockade and 'other operations by land, sea and air forces'. In many cases it may be appropriate to use these various measures in chronological sequence, only moving up the ladder as and when softer approaches have been tried and failed. But in other instances it may be that to go in early and hard, albeit on a limited scale, might avert much bloodshed and suffering. For instance it has been argued that had the UN, led by the United States, committed ground troops with air support in former Yugoslavia at a much earlier stage (e.g. to prevent the destruction of Vukovar in Croatia by the Serbs in 1991) this would have nipped civil war in the bud. This implies a judgement at the outset that gentler methods are bound to fail and that recourse to the methods of last resort (albeit under the general rubric of minimum force) were better taken earlier than later. The term 'last resort' need not, on this reading, be understood chronologically. There is much to be said for this view, but the necessary decision-making will always be very difficult.

Related to these principles is the question whether, in order to justify intervention, some *national* good for the intervening nations must be at stake. If you think not then how do you, as a politician, justify sending young British men and women to suffer and to die where *no* British interest is involved? And when things get rough, the television coverage harrowing, how highly do you rate your chances of sweating it out? Do you argue that if hard-line nationalists get away with murder in Bosnia then it will be the turn of Kosovo next followed by Macedonia and then perhaps a wider Balkan war; or at least the encouragement of villains in other places from which in the end Britain will suffer? This used to be a popular view and known as 'domino theory', but is now largely discredited. According to the British doctrine of Military Aid to the Civil Power, action should always address the 'here-and-now'; not what may possibly happen elsewhere in the future. Or do you contend, following Donne, that no man is an island? One man's death diminishes us all. In Henry Kissinger s words: 'humanitarian intervention asserts that moral and humane concerns are so much a part of American life that not only treasure but lives must be risked to vindicate them: in their absence American life would have lost some meaning'.[4] He adds that no other nation has ever put forward such a set of propositions: remembering the expansion of the Roman Empire I am not so sure about the latter point. But the trouble with this argument is that it proves too much. Why is there no thought of intervening in Algeria, Myanmar, Indonesia, Iran, Kashmir, Laos, the Lebanon, Peru/Ecuador, the Philippines, Sri Lanka? Will it do, morally, to say that these are not on our doorstep or, more to the point, our television screens? Or is Joseph Nye nearer the mark when he says 'a foreign policy of armed multilateral intervention to right all such wrongs would be another source of enormous disorder'.[5] Obviously, each instance has to be judged on its merits, balancing the destruction to be expected against the good one hopes to do by it.

Intervention

Trenchant criticism has been mounted against the whole concept of 'humanitarian intervention on strictly pragmatic grounds'. Philippe Biberson, president of *Médècins sans Frontières* put the general question 'whether generous intentions have been made to prolong and magnify the human distress that they were meant to ease. Lives are saved, but is it only to advance evil purposes, to sustain fear and hatred and the wars that they are used to promote.' He was speaking in the first instance of Rwanda, where the massive exodus of refugees was to a large extent planned and led by those who had just organised the massacre of perhaps half a million people. These leaders knew that they could rely on the international community to respond with material assistance and that there would be few, if any, efforts to isolate them from the mass of the refugees. They proceeded to control distribution of donated supplies, siphoning off large amounts, using violence and threats, preparing forces for a renewed offensive in Rwanda. Relief agencies walked straight into the trap. 'Contributing to this enroling of civilians by force with silence empties humanitarian work of its sense,' Dr Biberson says.

Similar criticisms might be made of the UN intervention in

Somalia. It certainly saved lives in some areas. But a new report from Mogadishu quotes one resident as saying, 'It is as if nobody came. It cost billions and we see nothing for it.' Worse than that, non-governmental organisations (NGOs) tolerated diversion and extortion and ended up paying hundreds of thousand of dollars to factional leaders. This played a critical role in enabling them to maintain their militias. Now that the UN has left, what else is to show for their efforts?

In former Yugoslavia, UNPROFOR certainly achieved admirable results in distributing food and other supplies to areas where they were desperately needed, but at the price of nearly 70 killed and hundreds wounded. Sceptics argue that UNPROFOR got convoys and airlifts through, and shelling of cities interrupted, only when the attackers had acquiesced for their own economic, political and military purposes. UN troops speak bitterly of their humiliation when they are not allowed to break barricades or return fire, even to resist being held hostage. Sarajevo survives but the war goes on. 'We are kept alive today to die on a full stomach tomorrow.'

Billions of dollars have been spent (the criticism goes) on alleviating humanitarian emergencies so that the better off parts of the world can feel that they are showing a modicum of compassion. This is eating into the capacity to provide promised aid for longer-term and more fundamental programmes. The UN Development Programme's administrator reports that emergencies, which consumed 25 per cent of UN resources in 1988, took 45 per cent in 1992. Perhaps a totally fresh approach is needed.

An alternative model of intervention is supplied by the type of operations which countries carry out unilaterally, normally in their own *backyard* (widely defined), in which the classical UN peacekeeping principles of consent, impartiality and minimum force are not merely abridged, with all the straining of language and ambiguity that this implies, but simply treated as inapplicable. This might be called the policing model, though it is important not to be prejudiced by overtones of oppression and arbitrary power. What is at issue here is the most economical method of controlling ethnic strife in terms of *ending the killing as quickly as possible with the least loss of life*. It can be argued that such operations, judged by this simple criterion, *sometimes* work better. Obvious examples are operations carried out in the past thirty years by France, the USA and Russia: being three states notoriously prone to interfere in the internal disputes of other lands.

The French make no bones about it. The 1995 *Loi-de-Programme* sets out a five-year rolling programme, designed to make the French armed forces more mobile, improving power-projection and sizing them to undertake three types of operation simultaneously: a regional conflict as part of a coalition, intervention in an overseas low-intensity conflict and limited UN peacekeeping operations. The French have a specialist non-conscript intervention force in the shape of the Foreign Legion. It is to France's credit that she comes second only to Pakistan in the number of troops committed to UN activities, 7,094 in 1995, spring, 10 per cent of the total; and her share of the peacekeeping budget is 7.6 per cent. Besides sending the largest contingent to UNPROFOR in former Yugoslavia, France also sent some 1,500 men each to Somalia and Cambodia. But France's actions on a national basis have been equally notable. Everyone knows that in the 1940s and 1950s there were disastrous campaigns in Indo-China and Algeria. Less well remembered is the fact that since 1962, when French troops were sent to Dakar to help President Léopold Sédar Senghor of Senegal maintain order after a coup attempt, France has engaged in more than a dozen major interventions in Africa including forays into Gabon, Chad, Zaire, Central African Republic, Togo and Comoros. Most were aimed at keeping in power leaders friendly to Paris. Sometimes it was to evacuate Europeans from danger as when French and Belgian parachutists intervened in 1978 in the Shaba Province of Zaire. Sometimes it was to remove a ruler, like Jean Bedel Bokassa of the Central African Republic, whose excesses had made him too much of a liability.

France carries off these manoeuvres with élan: no apologies, no qualms over sovereignty, the helicopters swoop in as though by right. France's prestige among African nations rises. In western capitals France is taken seriously again. And in Paris there is explaining to do only when the government *fails* to intervene, if French interests are seen to be at stake. The key seems to lie in the nature of the links between France and its former colonies. John Darnton, reflecting on this in the *New York Times*, says that France, and the large area of Africa which has twice as many French speakers as France itself, are held together by ties of language, love of French culture and the conviction that Paris is the centre of the universe.[6] It is a Faustian bargain. Let in French technocrats to run state enterprises, trade mainly with the mother country, sign a military assistance pact and you will be looked after. France will prop up your economy by giving you the African Franc (supported by the French treasury although recently devalued) and rush its army to your side if trouble comes. When killing erupted in Rwanda in April 1994 governments quickly pulled out the 5,500 UN troops serving there. The French then sent a sizeable force of soldiers to establish a safe area for refugees in the south west of the country, thereby saving many lives.

The United States of America is now regarded as the world's natural hegemony: only where it leads will the world follow—so it is said. The Gulf War of 1990–91 is the classic instance. But we are talking here of interventions in the internal affairs of strife-torn states particularly in one's own *backyard*. In theory the USA renounced any right of unilateral intervention in the western hemisphere by the 1947 Rio Pact: armed attack on any American nation should be met by *collective* action. But in practice the US has continued to intervene on its own when it sees fit. In October 1983 the US occupied Grenada (which was technically British) losing some 19 personnel. In December 1989 they invaded Panama to oust the dictator Noriega, and lost 23. The invasion of Haiti in autumn 1994 is exactly of this pattern. But in this case the US had the backing of a UN Resolution and other nations forces have now replaced a large part of the US force.

Outside its own backyard the US has never found its stride. Vietnam was, of course, the great defining trauma. The Lebanon was another failure. The reasons for going into Somalia in December 1992 were wholly humanitarian—people were dying in hundreds from hunger and banditry—and to begin with much good was done, food convoys got through, lives were saved. The idea was to go on from saving lives to reconstructing the country and soon the US troops became subsumed in a UN operation. In this phase the US reverted to its backyard mode, threw impartiality out of the window, demonised one faction in the contest and went all out to 'get' General Aideed. Predictably this failed; 18 Americans died in a raid and Congress foreclosed. It is important to recognise what went wrong. It was not simply that the abandonment of impartiality fitted ill within a UN-led operation. The crucial point was the failure of the Americans in this instance, to apply the principles of low intensity operations. Notable among these are a clear political objective, an efficient intelligence network, the closest coordination between civil, military and police authorities and weaning the 'hearts and minds' of the population away from the insurgents in favour of the legitimate authorities. None of this happened in Mogadishu, where forces (rangers, gunships) were used indiscriminately and in such a way as to subvert the purposes of the operation as a whole. Trying to make sense of it all the Americans have come up with Presidential Decision Directive (PDD) No.25, based on the two premises that the UN is an inadequate instrument of peacekeeping and is costing the US too much. In fact the US share of UN peacekeeping costs is 31 per cent, although they are about $1 billion in arrears. But their contribution to UN operations has always been thin and reluctant: in the 1990s it was mainly to UNPROFOR, but with a handful of observers in Angola and West Sahara. The criteria for taking part, set out in the PDD, seem to emerge from answering the following questions:

- Would UN involvement advance US interests?
- Is there a condition of international aggression, humanitarian disaster, disruption of established democracy or gross violation of human rights?
- Are the objectives and mission clearly defined in scope and duration?
- Are realistic criteria in place for ending the operation?
- Is inaction considered unacceptable [a good double negative]?

If all these are met a further list of considerations comes into play: what is the risk to US personnel? Is their participation essential to success? Are the command and control arrangements acceptable? Is there a prospect of public and congressional support? Is there a plan, a strategy that integrates political and military goals, sufficient resources, a commitment to a decisive outcome? Decisions, they say, will rest on the cumulative weight of all these factors. Fine. But what this means in practice is that, where ethnic and tribal conflicts are concerned outside their own backyard, American strategy is now required to be as far as possible *risk-free*. On no account commit ground troops. Support the weaker side by pumping in more arms. Use air power—on an Alliance basis if possible—fairly ruthlessly, and let existing classic UN-type operations go hang. I hope this doctrine will not prevail in Bosnia. As General Michael Rose pungently remarked:

'Patience, persistence and pressure is how you conduct a peacekeeping mission. Bombing is a last resort because you cross the Mogadishu line. If someone wants to fight a war here on moral and political grounds, fine, great, but count us out. Hitting one tank is peacekeeping. Hitting infrastructure, command and control, logistics, that is war, and I am not going to fight a war with white-painted tanks.'[7]

Turning finally to the Russians. Their empire is by far the oldest, dating back 1,000 years. Yeltsin effectively disbanded the Russian empire at the end of 1991; now the Russians are rowing back, understandably. So far as the UN is concerned, the Russians have behaved relatively well: not using their veto on occasions when they might have wanted to; sending troops to UNPROFOR; and in the 'Contact Group' for Bosnia exercising their influence in a generally positive way. It is in their conduct towards their near abroad—that is, towards the republics of the old Soviet Union—that there has been concern. Russian interventions in the newly independent republics of the former Soviet Union have been of three types:

- Peace-making: as in Tajikistan where Russian troops (who had never withdrawn) and other CIS forces (notably Uzbeks) intervened on one side in the civil war, thus ensuring victory for the forces of their chosen champion Islam Karimov against the democratic-Islamic opposition; since when Russia has sought CIS partners in what comes closer to a UN type peacekeeping operation.
- Province-making: as in Moldova where the former Soviet 14th Army (which likewise stayed put) supported and possibly incited separatists to the East of the Dneister to break away from control of the Moldovan government and thus effectively blackmailed Moldova into joining the CIS.
- Peace-preventing: as in the Caucasus, where Russia intervened on *both* sides in two different disputes; helping first the Abkhaz separatists and then the Shevardnadze government in Georgia; and first Armenia and then Azerbaijan in Nagorny-Karabakh. The result in both cases is that the dispute has simmered on while both Azerbaijan and Georgia have been forced to accept membership of the CIS and the basing of Russian troops in their countries.[8]

A number of motives seem to be in play. First Russian strategists argue that the Caucasus remains important to Russia as its best access to the Middle East and Moldova to the Balkans. Second, Russia likes the idea of leaving border troops along as many of the old frontiers as possible, particularly on the southern approaches, if only to keep others from getting a foothold there. Third, there is the practical point that keeping military bases in these places saves adding to the massive problems of re-settling troops from Germany and Eastern Europe. But there is more to it than that. Russia has, it thinks, a legitimate concern for the 26 million ethnic Russians now living in the other republics. As an ex-superpower it feels she has a *right* to influence events in her 'near-abroad'.

The trouble is that, given Russia's desire to fend for its ethnic population outside Russia, sanctioning such efforts under the rubric of protecting 'human rights' would give it virtually *carte blanche* within the whole of the former Soviet Union. That would be going too far. But there is some justice in what Russia is proposing and it is wrong to treat the Russian attitude as wholly self-serving. The need is to encourage a more adult approach based on the principles of international law and good practice: the need for consent freely given, for transparency, for reversibility if consent is withdrawn. The West needs to ask how it can help the smaller states to integrate themselves better into the world community and so enjoy a better bargaining position. There is no denying that what has happened in Chechnya has set this cause back. But there is also need to recognise that Russia, in setting aside the classic UN Peacekeeping principles of consent, impartiality and minimum force, is acting no differently from other powers *in their own back yard*: the US in the Caribbean, the French in Francophone Africa. Arguably the results, in terms of stopping the killing, have sometimes been better than the UN has achieved by more consensual methods in e.g. Angola, Somalia or ex-Yugoslavia.

It has been the burden of this chapter that, setting the political background of the old imperial nations aside, some of the principles developed for dealing with internal conflicts—under the rubric of Low Intensity Operations—may better serve the ends of humanity than the methods presently being attempted by the international community in the form of 'Wider Peacekeeping'. But once the principle of impartiality is set aside a further crucial question has to be confronted. It is an issue of principle which has underlain much of the discussion about former Yugoslavia and deserves now to be brought out into the open. Should the UN, in its conflict-resolution role, be prepared sometimes to support one side or the other rather than remain scrupulously impartial? This in turn raises other questions. Might it be right to support the stronger side on the grounds that this will enable the war to be brought to an end as cheaply in human lives and as quickly as possible? Or should the UN espouse the weaker side, on the principle of solidarity with the victims? The latter approach might well seem to have justice on its side—to be the more moral/ethical way of setting about things. But it might make for a longer war. And intervention on either side would be an overtly political act. Would the UN be prepared to take this path given its dependence on donor nations to provide troops and resources, tolerance by 'host' governments and favourable publicity? Bosnia has provided a yet more difficult case because of Russia's Orthodox and Slavic links with the Serbs, Germany's Catholic links with the Croats and almost the whole Islamic world with the Bosnian Government. The prospect of a rift along these lines within the Security Council could create a situation fraught with difficulty and dangers. It can be argued that no international body (be it the UN, OSCE, EU, ECOWAS or whatever) could ever agree to support one side against the others in a conflict of this kind. From here it is a short step to arguing that if one is not prepared to take tough decisions of this kind it is better not to intervene at all, but rather to rely instead on the various non-military mechanisms referred to at the outset. But when all these have already been tried (or not applied) and failed: where killing is in progress and genocide is both flagrant and continuing, to rule out a military response *a priori* is a counsel of despair.

A possible way out of this dilemma is to rely not so much on the UN *per se* as on a 'coalition of the willing'. On this basis it may be possible to take the necessary tough and early decisions, to intervene decisively in an internal conflict where the Security Council is powerless to act, or to act quickly or decisively enough. This suggestion in turn raises further problems:

- How much regard should be paid to legality? In some cases there might be a request for help from an established government (e.g. Bosnia) but obviously not always.
- How many of the 'willing' will there be?
- What happens if they intervene on the 'wrong' side (as arguably the French in Rwanda)?
- What if two lots of the 'willing' intervene on opposite sides (*vide* the Cold War)?
- What happens if the intervention fails to achieve the quick and

decisive results hoped for? This is a predictable, indeed normal outcome. Should the intervening forces 'declare success' and leave or should they reinforce failure thus entering the quagmire (Vietnam syndrome)?

- And finally, given early success in the sense of disarming the trouble-makers and suppressing any who fail to co-operate (as in Haiti), what should follow? There may then be a need to set up an interim administration, organise elections and generally engage in 'nation-building'. Faced with a possible long haul of this kind how many will be 'willing' to begin?

A role for Britain?

The question remains what have the British to offer towards a new approach on how coalitions should act in support of peace-making. The British like most other nations have laid great stress in the reorganisation of their armed forces, on rapid reaction and the ability to despatch forces overseas. In manpower contributions to the UN Britain came fourth in the league in 1995, with 3880 troops committed, lagging only Pakistan, France and India. But the British act quite differently from the French in dealing with ex-colonies. Once they have handed over power their military involvement is confined to the provision of advisers and training teams, sending troops on exercises (often with a civil relations slant), and sales of military equipment. And yet, in the process of de-colonisation, the British had as much experience of dealing civil conflict as any nation. Looking back over 50 years, and disregarding the four actual wars—Korea, Suez, Falklands and the Gulf—what the British have actually been doing is coping continuously with ethnic, tribal or religious struggle; only in 1968 were no British troops killed in action anywhere. Sometimes they have attempted to cope according to a *peacekeeping* mandate, receiving brickbats from both sides. Very often, as in the cases of Palestine and Aden, the results were most unhappy: the British left without finding any lasting resolution to the underlying conflicts, which persist to this day. Elsewhere their role was quite different, defeating rebellions against the colonial authority: the Mau Mau in Kenya, a communist-led insurgency in Malaya and revolt in the Oman. This led the British to develop very sophisticated and well-tried doctrines, equipment and training to deal with these types of operation known collectively as Low Intensity

Operations, or Counter-Insurgency. Tragically one of the longest-running and latest instances has been in Northern Ireland.

The answer to the present question is therefore simple. Without in any way seeking to disparage the British Army's new *Field Manual* 'Wider Peacekeeping', the British need also to recognise, and be prepared to implement, the type of doctrinal approach at which they have long excelled in the peace-*enforcement* context. The forces exist. British infantry, including the parachute regiment, the Commandos and, where appropriate, the Gurkhas, are the equals of any to be found elsewhere in the world in terms of professionalism, toughness, urban, mountain and jungle skills, restraint and the trick of appealing to the hearts and minds of the inhabitants. The Manuals are still extant, Staff College notes and exercises barely cold. The training facilities are still in hand. All this can be offered to other nations who wish to learn from us.

To put the point at its most cynical, if the troubles in Northern Ireland are truly at an end for this generation (which God grant) then what else will the British infantry be for? In classical peacekeeping there are many countries who can provide admirable foot-soldiers, and in far greater numbers than we can. In the realm of communications and logistics the British have an edge over most other countries and their contribution is valued accordingly. But in counter-insurgency, and in enforcement action British experience is more or less unique. In offering such expertise the British should keep their nerve and even, where appropriate, be prepared to take a lead.

PART 5

ADAPTING THE COLD WAR ARMS CONTROL REGIME

13 CONVENTIONAL ARMS CONTROL REGIMES FOR POST-COLD WAR EUROPE

Jane M.O. Sharp

In 1992 two agreements entered into force which imposed numerical limits on five categories of heavy military equipment, and on military personnel, in NATO and the states of the former Warsaw Treaty Organisation (WTO). These are the Treaty on Conventional Armed Forces in Europe (CFE) which sets equal limits for NATO and the WTO for main battle tanks, armoured combat vehicles (ACVs), artillery above 100mm calibre, attack helicopters and combat aircraft; and the Concluding Act of the Negotiation on Personnel Strength of Conventional Armed Forces in Europe which sets national manpower limits on all the CFE state parties. In addition, all 52 states in the Conference of Security and Cooperation in Europe (CSCE) agreed in the Vienna Document 1992 to abide by a set of Confidence and Security Building Measures (CSBMs). Together these agreements, which provide for elaborate exchanges of information and regular inspections as well as numerical limits, constitute a regime that many hoped would become an integral part of a pan-European security system under the auspices of the CSCE.

These hopes were not realised, not least because former Yugoslavia is not a party to CFE and its constituent republics ignored their commitments to the CSBM regime. Moreover, while CFE in general has a salutary effect on force levels throughout western Europe, it also triggered the destabilising NATO practice of 'cascading' excess treaty limited equipment (TLE) from Europe's stable north-west to Greece and Turkey in the volatile south-east.[1] The biggest problem, however, was that Russia claimed that the CFE Treaty discriminated against the Soviet States. At the VE celebrations in Moscow in May 1995, President Yeltsin told President Clinton that Russia had fallen into a trap when it inherited CFE which was signed by the Soviet Union under completely different political and military circumstances.[2] Thus while all the parties to CFE complied with their overall ceilings, Russia and some of the CIS states refused to comply with Article V which set special sub-limits in the so-called Flank Zone.

This chapter argues that in the context of shaping a stable post-Cold War Europe, Britain should take the lead in adjusting the current conventional arms control regime to the new realities. In the long term this could mean bringing all the European countries (that is the former neutral and non-aligned countries including all the states that used to make up Yugoslavia) into the same arms control regime. In the short term, however—say the first term of the next government—the most important task.will be to respond to Russian requests to revise the Treaty on Conventional Forces in Europe (CFE), not least to make an eastward enlargement of NATO more palatable to the Russian military.

This will not be easy because, since the collapse of the USSR, western diplomacy towards Russia has been incoherent at best. In contrast to the western effort to transform Germany from totalitarianism to democracy after the Second World War, the major powers made no attempt to set down markers for Russian behaviour after the Cold War. On the contrary, western governments were extraordinarily permissive towards gross contraventions of international norms, such as the Russian bombing of civilians in Grozny, but unreasonably inflexible in responding to what many western analysts believe are legitimate Russians concerns about CFE.

To be permissive where condemnation is in order, and rigid where flexibility is appropriate, does Russia no favours and certainly does nothing to further the process of reform there. Nor does inconsistent and *ad hoc* diplomacy serve the interests of the western security community. The lack of a western response to Russian grievances about CFE is especially puzzling because Russian diplomats and senior military officers raised the issue in the proper channels, according to procedures laid down in the Treaty.

CFE limits were well designed for a bipolar Europe and for the detente period that ended the Cold War, but from the Russian perspective have now been overtaken by events. The treaty was negotiable in 1989–90 largely because when Mikhail Gorbachev assumed the leadership of the Soviet Union in the late 1980s he adopted a conciliatory policy towards the NATO countries. Specifically, in contrast to his predecessors, he was willing to open up Soviet territory to inspection, to make unilateral withdrawals of Soviet troops from eastern Europe, to make unilateral cuts in those categories of equipment in which the Soviet Union enjoyed superiority and in general to make whatever concessions the west demanded. As Gorbachev's conciliatory attitude built confidence and trust between the west and the Soviet leadership, however, it humiliated the Soviet General Staff, not least because the initial western responses were so meagre.

Between treaty signature in 1990 and the end of the reduction period in 1995 Russian foreign policy and military doctrine changed radically from that of the Gorbachev leadership that negotiated the CFE treaty during 1989–90. Boris Yeltsin may have looked like a democrat atop the tank in front of the Russian White House in August 1991, but in 1995 he no longer appeared to believe in the conciliatory pro-western policies of Mikhail Gorbachev.

Senior Soviet military officers complained about the terms of CFE from the day it was signed.[3] They accused Soviet foreign minister Eduard Shevardnadze of selling out to western interests and took several steps to evade the treaty provisions. These included the transfer of TLE outside the zone of application and the redefinition of ground troops as naval infantry and coastal defence units. These actions irritated the other Treaty partners and delayed ratification. As John Major told the House of Commons, there was little inclination to permit the Soviet navy more tanks than the British Army.[4]

Eventually, in May 1991 Gorbachev insisted that the Soviet General Staff adhere to the terms of the treaty as signed in November 1990. Thus on 14 June 1991 the Soviet representative at the Joint Consultation Group (JCG), the group specifically established to resolve compliance ambiguities, agreed that TLE deployed with coastal defence and naval infantry units would count in the national CFE ceilings, and promised to account for the equipment removed from the CFE zone before treaty signature.

Ratification was further delayed by the collapse of the USSR after which new negotiations were necessary to re-allocate former Soviet military assets among the newly independent republics. This further alienated the Russian military from the treaty as General Pavel Grachev warned US Secretary of Defence Les Aspin in March 1993. In September 1993 President Yeltsin joined the military in opposing CFE and wrote to several western leaders requesting revisions. Russia also lodged formal complaints to the other CFE parties at the (JCG) in Vienna. Russia asked the other state parties to convene an Extraordinary Conference to consider their request for revisions as provided for in the treaty (Article XXI.2). The other CFE states responded by saying that Treaty revisions should await the review conference scheduled for May 1996, six months after the end of the reduction period ended in November 1995.

The Russian case against CFE

The Russian case for treaty revisions rests on three arguments:

- The military and political environment in Europe has radically changed since the treaty was signed, making CFE limits increasingly discriminatory for Russia.

Since CFE was signed in November 1990, Russian power (political, economic and military) has diminished relative to that of the western democracies. The first political setback occurred before CFE was signed, in July 1990, when the German Democratic Republic, Russia's main prize in the Second World War, effectively joined NATO (and the European Communities) via unification with the FRG. The following year the Warsaw Treaty Organisation (WTO) collapsed and the USSR devolved into its constituent republics, leaving 25 million Russian speakers outside the Russian Federation. All Russia's former allies in central Europe sought to enhance their political, economic and military security by trying to join western institutions: the European Union, the Western European Union and NATO (see Chapter 9). At the JCG in Vienna in July 1994, Russian delegate Viatcheslav Kulebyakin asserted that NATO expansion would violate CFE. 'Whether we like it or not, he said, 'the treaty is a bloc to bloc agreement.' In October defence minister Pavel Grachev repeated these complaints, as did General Vladimir Zhurbenko at a December press conference at the Ministry of Defence in Moscow.[5] Central European states disputed this interpretation since they adhered to the treaty as individual states and not as members of the Warsaw Treaty Organisation, then manifestly in its death throes.[6]

In early 1995, as NATO moved ahead with its study of enlargement, Russian military spokesmen continued to assert that incorporating former WTO states into NATO would destroy the treaty.[7] Responding earlier and more positively to Russian requests to revise CFE would not only have preserved the integrity of the treaty, but would also have made NATO enlargement more palatable to Moscow. In a 14 March 1995 interview the chairman of the Duma Foreign Affairs Committee, Vladimir Lukin, acknowledged that a deal on CFE flank limits would certainly facilitate Russian acceptance of NATO expansion. As Lukin put it: 'If Russia is allowed to deploy its forces as it wishes then NATO approaching our borders will not be so dangerous'.[8] Not many weeks afterwards, however, at a conference at the Kennedy School of Government at Harvard, Lukin reversed the linkage, asserting that if NATO expanded he would personally see to it that Russia did not comply with CFE. NATO expansion need not undermine CFE, however, if with each new state's accession the other NATO allies adjust their CFE ceilings downwards to prevent any increase in the NATO group ceiling.

- CFE destruction requirements are too expensive, and the time allowed for equipment reductions too short for Russia and the other Slavic states, which have greater reduction obligations than other state parties.

To meet these concerns the JCG made five revisions to approved destruction procedures during 1993–94. The first was in May 1993 with respect to tanks and ACVs, the second in June 1993 with respect to ACVs, the third and fourth in October 1993 further adjusted techniques for conversion of tanks and ACVs to non-military use, and the fifth in March 1994 adjusted procedures for destroying artillery.

The treaty provides for the destruction of equipment (TLE) in excess of treaty limits in three stages. The first two reduction periods were successful in that each of the Groups of States (NATO on the one hand and the former WTO states on the other) achieved the treaty targets of 25 per cent reductions in the first year and 60 per cent in the second. In the former WTO group, however, several of the former Soviet states were not in compliance with respect to their national ceilings. In February 1995, President Alexander Lukashenko of Belarus ordered a halt to all CFE destruction procedures on the grounds there was no money left to pay the workers at the destruction plants. Lukashenko made his announcement on 17 February in a speech to a Special Forces Unit in Pukhovich. A few days later President Yeltsin came to Minsk to sign an agreement establishing two Russian military bases in Belarus and arranging for Russian troops to patrol the border between Belarus and Poland.[9] Immediately after the Yeltsin visit, at a dinner in Minsk marking Red Army Day on 23 February, Lukashenko gave as an additional reason not to destroy the TLE the need to be on guard against the expansion of NATO; specifically voicing concern about Poland and Lithuania joining the western alliance. NATO expansion and financial constraints were again invoked as reasons to halt TLE destruction at ceremonies marking the holiday on 24 February.

In official messages informing other CFE state parties, Lukashenko invoked only financial constraints as a reason to halt destructions and stressed that Belarus did not want to abandon the treaty. German foreign minister Klaus Kinkel chastised Belarus, claiming the treaty would not survive unilateral abrogations of this kind.[10] More pragmatically, Britain offered technical assistance and tried to persuade officials in Minsk to recommit to the reduction schedule. The US sent equipment valued at $3 million, but this was not much help as Belarus had neither wages for workers nor fuel for the donated machinery.

- Article V sublimits discriminate unfairly against Russia.

During the CFE negotiations NATO's main purpose was to curb Soviet conventional forces and thereby reduce the likelihood of a military clash between NATO and the WTO in central Europe. An important organising principle was to impose limits on heavy equipment in roughly concentric bands with the most severe limits in the central zones. Turkey and Norway, however, demanded curbs on the amount of Soviet equipment in the designated Flank Zone on their borders. The Soviet General Staff also modified the original zones, redefining the Kiev Military District as part of the Central rather than the Flank Zone, in order to accommodate troops and equipment returning from Afghanistan. These changes effectively cut the levels of equipment permitted in the Flanks, and left large quantities of the most modern Soviet heavy equipment on Ukrainian territory.

Article V stipulates that each CFE Group of States (NATO comprises the western Group and the former WTO states comprise the eastern group) may hold no more than 4,700 tanks, 6,000 artillery pieces and 5,900 armoured combat vehicles (ACVs) in the area designated as the Flank Zone. For the eastern Group of States this included Romania, Bulgaria and the Leningrad, Odessa, Transcaucasus and North Caucasus Military Districts of the USSR. Before CFE was signed in November 1990, within the WTO Group the Soviet Union was allocated 3,850 tanks, 4,575 artillery pieces and 3,400 ACVs in the flank zone. When the USSR collapsed into its constituent republics six new states had to share the former Soviet allocation. New allocations were agreed by all the former Soviet states at Tashkent in May 1992, when Russia's allocation for the Flank zone was reduced to: 700 tanks, 1,280 artillery pieces and 580 ACVs.

Solutions proposed by Russia

In September/October 1993, invoking Article XX which allows amendments to the treaty, Russia called for either a redrawing of the CFE zones, or suspension of Article V. General Vladimir Zhurbenko, First Deputy Chief of the Russian Armed Forces, asserted that forces would be deployed to trouble spots in the Caucasus and elsewhere even if this contravened the treaty.

In February 1994, at the CFE Joint Consultative Group in Vienna, Zhurbenko referred back to the June 1991 unilateral statements in which the USSR undertook specific obligations to limit equipment deployed with the naval infantry and coastal defence units, suggesting these restrictions now be lifted. This was repeated in late April, when Russia also requested a modification in TLE storage provisions. Specifically Russia wanted to suspend paragraphs 9 and 10 of Article X which capped the total amount of equipment that could be removed from storage by each group and the length of time the equipment could remain out of storage.

Exasperated by the lack of any western response to all these requests Russia threatened (a) to withdraw from the treaty, and (b) not join NATO's Partnership for Peace unless CFE was revised. Since then Russia has signed up to PfP, but the threat to withdraw from the treaty still stands.

In September 1994, General Grachev spelled out Russian equipment needs in the Flank, most of which he would deploy in the North Caucasus Military District: 1,100 tanks (allowed CFE ceiling in is 700); 3,000 ACVs (allowed CFE ceiling is 580) and 2,100 artillery pieces (allowed CFE ceiling is 1,280). The Russians clearly did not intend to reduce their Flank holdings to the CFE-imposed ceilings.

In November 1994 at the Preparatory Committee meeting for the CSCE summit in Budapest, Russian delegates asserted that the full membership of CSCE should debate CFE revisions at the summit in December. They failed to get the other parties to agree, however. In Budapest Russia also tried to gain endorsement for 'third party peacekeeping' that is, for Russia to settle disputes between two of its neighbours under CSCE auspices. Russia hoped that a CSCE seal of approval of such 'peacekeeping' in the CFE zone of application would exempt those forces and their equipment from CFE limits.

During the first half of 1995, parallel to the war in Chechnya in which Russia lost at least 2,000 Russian soldiers, Russia pressed even harder on the CFE issue. In April General Vladimir Semenov announced that by 1 June the 58th Russian army with 2,500 tanks would be permanently deployed in the North Caucasus Military District; implied contravention of Russian Flank limits if normal levels of equipment were deployed. Kozyrev proposed new Flank limits to Warren Christopher to accommodate the 58th army, but Christopher said that western governments would not consider any changes before the end of the treaty reduction period in November 1995; that the correct time to discuss changes was at the Review Conference scheduled for May 1996. This line was repeated at the VE celebrations in Moscow in May 1995 when President Clinton said he had some sympathy for Russia's position on CFE and promised that the US would look sympathetically at treaty revisions at the May 1996 conference, but not before.

Attitudes of the other state parties towards treaty revision

Article XX allows for amendments to the treaty. Technically it would not be difficult to either redefine the zones or suspend Article V limits. Politically, however, all the state parties must approve any changes. Given the laborious negotiations on Flank limits both prior to signature in 1990 and again after devolution of the Soviet Union, there was little interest among the other CFE parties in opening up the treaty for fear the entire structure could collapse, and considerable resistance to taking steps which appeared to condone Russia's brutality in Chechnya.

Central European attitudes

Having suffered Soviet domination in the Warsaw Pact throughout the Cold War, and Russian high-handedness in international forums during the 1990s, central Europeans in general were not in favour of granting Moscow any special concessions at CFE. On the other hand, if the treaty opens up, the central Europeans may want changes of their own.[11]

Czechoslovakia

Czech delegates were more sympathetic to the Ukrainian than to the Russian request for Flank revisions. At the JCG in late March

1994 the Czech delegate said that Ukraine faced serious threats to its security and deserved higher ceilings, whereas Russia was threatened by no other state and enjoyed military superiority over all her neighbours.[12]

Poland

Poland stood firm with the general NATO view that Russia should comply with the treaty in full and wait until the May 1996 conference for any revisions. Nevertheless, while fearful of the treaty unravelling, Poland was also interested in revising the storage provisions in the treaty and in changes that might lessen the heavy military presence in the Russian Oblast of Kaliningrad, estimated in April 1994 at more than 100,000 troops plus their equipment.

Hungary

Hungary also opposed any treaty revisions before November 1995, but accepted the possibility of amendments at the May 1996 conference. Hungarian officials are acutely sensitive to the dangers of great power deals in multilateral negotiations and worried that the US would again pander to the Russians and make a deal before November 1995. Hungarian delegates to CFE were unhappy with the James Baker–Eduard Shevardnadze deals in 1990 and the Lynn Hansen–Oleg Grinevksy deals in 1991. They feared something similar in 1995 that the other state parties would be more or less forced to accept.[13]

Bulgaria

Bulgaria, like most central European states valued the benefits of the CFE regime and was nervous of opening up the treaty for revisions. In November 1993, Dimitur Mitkov, an official in the Ministry of Defence in Sofia, said that he feared further instability and arms racing in the Balkans if treaty limits were open to revision.[14]

Romania

By contrast, Romania, never happy with its CFE allocations within the WTO Group, indicated in July 1995 that if Romania is not in the first intake of new NATO members it would leave the CFE treaty regime.[15]

Former Soviet states

Ukraine raised the Flank issue at the JCG before the Yeltsin letter was sent to NATO leaders in mid-September 1993, and consistently supported the Russian call for revisions through 1993–95. Like Russia, Ukraine also proposed 'a partial modification of the obligations associated with the Declaration of the USSR Government of 14 June 1991'. That is, Ukraine wanted TLE with naval infantry and coastal defence forces to count only in the overall Group and national ceilings, but not in the Flank subceiling. This proposal was repeated again by Lt General Gennadi Gurin at the JCG in January 1995.[16] Gurin also notified the other state parties that, owing to the prohibitive cost of rehousing troops to comply with the Flank limits on TLE, Ukraine would continue to deploy troops and equipment above CFE limits in the former Odessa Military District. Gurin said the legal basis for this continued deployment could be found in Article V para 1 subpara (C), which under certain circumstances permits the temporary deployment of 153 battle tanks, 241 ACVs and 140 artillery pieces above national limits.

Because of the way the CFE zones were drawn up in 1990 Ukraine ended up holding large amounts of modern Soviet equipment which it then had to pay to destroy in compliance with CFE. Ukraine repeatedly complained about the costs of its heavy reduction and destruction obligations under the treaty. Nevertheless, in March 1994 Ambassador Yu Koshenko said that the May 1996 CFE Review Conference should seek lower tank ceilings and impose limits on naval aircraft as current levels were far too high.[17]

Of the other former Soviet CFE parties, Armenia and Belarus, consistent with their political alignment, both support Russian claims for revisions. Azerbaijan denied Russian requests for revisions as it did Russian requests for some of Azerbaijan's CFE allocations. Moldova and Georgia by contrast appear to have caved in to Russian pressure to give up some of their CFE allocations.

NATO attitudes to CFE revisions

The initial NATO response to President Yeltsin in September 1993 was to resist any revision of the treaty, but to continue debate in the JCG. There was little interest among the other European allies either to renegotiate Flank limits or turn a blind eye to Russian treaty violations. NATO governments argued that CFE should first be fully implemented before state parties began to negotiate amendments, and that the time to consider revisions would be after November 1995, at the Review Conference already scheduled for May 1996. They feared any revisions before November 1995 could prompt other requests and risk unravelling the whole treaty.

While no NATO state party relished the idea of renegotiating any part of the CFE treaty, the degree of opposition to Russia's proposal to suspend Article V varied according to their distance from Russia. The NATO Flank states closest to Russia (Turkey and Norway) being most opposed, and those furthest away (the USA and the UK) being most relaxed.

Norway was one of the states most anxious to limit Soviet equipment in the Flank Zone during the CFE negotiations 1989–1990, claiming that the treaty must adequately address the security of all European states not only the central zone.[18] Accordingly Norwegian delegates to the JCG consistently opposed Russian requests to revise Article V.[19]

Turkey, even more than Norway, was concerned about an increased Russian presence on its borders, and like Norway pressed hard during 1989–90 to limit Soviet equipment in the flank zone. Even before Russia formally asked for suspension of Article V in September 1993, Turkey had complained on several occasions that Russia risked abrogating CFE limits with its military interventions in the Transcaucasus.[20] Turkey feared Russian imperialistic ambitions in the Caucusus and was firmly convinced that Russia stirred up the trouble that erupted in the early 1990s in Georgia and fuelled the dispute between Armenia and Azerbaijan over Nagorny-Karabakh. At the June 1944 meeting of the North Atlantic Cooperation Council in Istanbul, Turkish delegates were especially angry when Russia tried to insert a reference to CFE revisions in the final communiqué of the meeting.

Turkey was in a weak position, though, to argue against special concessions for others given its own privileged position. Not only has Turkey (with Greece) benefited enormously from the modernisation of its heavy equipment from NATO's cascading of TLE from the north, but during 1989–90 Turkey managed to exclude the south-east corner of its own territory from CFE limits on the grounds that the region faced threats from states to the south that were not subject to similar restraints: namely Iran and Syria.

France opposed any changes primarily on the grounds that the whole edifice could crumble if the JCG gave in on the Article V issue. Of all the NATO countries France reacted most strongly to Russian brutality in Chechnya in early 1995 and claimed this was no time to reward Russia with special concessions in the CFE forum.[21]

In the early 1990s German policy-makers appeared to regard the overall bilateral relationship with Moscow as much too important to be sacrificed on the altar of CFE. Germany thus tried to be sensitive to Russian concerns about Flank restrictions without actually revising the treaty. In a paper presented to one of the CFE verification seminars, former CFE delegate, Ambassador Rudiger Hartmann, suggested that one way to deal with the Flank issue was to accept temporary deployments of Russian TLE in excess of CFE limits in the Flank Zone.[22] This was the solution adopted by Ukraine to solve its own predicament, but the size of temporary excess deployments allowed was too small to accommodate the extra TLEs demanded by Russia. Germany supported Russian requests for amendments to costly equipment destruction procedures. On the other hand on numerous occasions German foreign minister Kinkel stated that CFE parties could not abrogate the treaty unilaterally and expect it to survive.[23]

The US and UK governments were the most responsive to Russian concerns about CFE, even though in late 1993 and early 1994 they appeared as negative as their NATO partners to any revisions of the treaty. Despite all the arguments against tampering with the treaty before the end of the reduction period there were at least four arguments in favour of accommodating the Russians on Article V. The first was not to jeopardise the delicate Russian–American arrangements in place for the cooperative denuclearisation of former Soviet nuclear weapons. With Republican domination of the Congress after the November 1994 election President Clinton wanted to avoid a Russian breach of the CFE treaty which would automatically cut off the Nunn-Lugar funding set aside for denuclearisation. These funds were conditional on full Russian compliance with existing arms control agreements.

The second reason for flexibility to preserve CFE was that the treaty is more valuable to NATO than it is to Russia. Not only did CFE limits cut substantially into Russian and other former Soviet republics' holdings of heavy equipment, but the compliance regime produces valuable intelligence about the military force postures of all the former Soviet republics west of the Urals. Western military establishments were therefore more sympathetic than foreign ministries and parliaments to Russian requests to suspend the Flank Zones. They argued that the incremental forces Russia could thereby deploy were trivial compared to the cuts that have already been achieved since the treaty entered into force.

For Russia, which bore the heaviest reduction and destruction obligations, the treaty offers no such advantage. CFE hardly made any dent in NATO equipment levels and, even without the inspections and information exchanges mandated by the treaty, Russia had access to many open sources for information on the levels of western forces.

The third reason to have accommodated Russia on CFE earlier would have been to make NATO enlargement more acceptable to Moscow. Had NATO responded to the Russian request for an Extraordinary Conference to discuss Article V in September 1993, the debate over NATO enlargement might not have become so rancorous during 1994–96.

The fourth reason to accommodate Russia on CFE is to put some real meat on the bones of the new NATO–Russia partnership. The danger is that, like NACC and PfP, the NATO–Russian partnership could end up with no real agenda, and become yet another talking shop that neither side takes seriously.

The modalities of revisions

The western allies misjudged the importance of the CFE issue in 1993–95 and chose not to confront the inflexibility of Norway and Turkey on this issue. Rather than lobby for revisions, British and American delegates played for time (both at the JCG and privately) by pointing out to their Russian counterparts how much flexibility there was inherent in the treaty, for example:

- Russian troops in the North Caucasus could be issued with equipment that was not limited by CFE, e.g. trucks, infantry weapons, artillery below 100mm calibre and certain tracked personnel carriers.
- Unlike ground force equipment, treaty limited aircraft and helicopters could be moved from zone to zone.
- Article III exempts some equipment with internal security forces from CFE limits.

- Article V.1 (c) allows the temporary deployment of an extra motor rifle brigade to deal with an emergency crisis.
- Article XII (as amended in Oslo in 1992) allows the deployment of 100 ACVs (outside the treaty limits for ground forces) with internal security forces.
- Equipment for Russian units in the CFE zone could be stored outside the zone, but close enough for rapid deployment in an emergency.
- By agreement with other CFE parties in its Group of States, Russia might be able to negotiate adjustments in its national ceilings.

Russia claimed, however, that even if all these loopholes were exploited it would not solve their problems in the Caucasus. Britain and the US then set up a series of trilateral meetings in 1994 at the deputy foreign minister level to try to resolve Russian problems without abrogating the treaty.[24] In letters to Bill Clinton and John Major in December 1994 Boris Yeltsin complained that these trilaterals had not been productive and called for more serious efforts to meet Russia's interests, including a redefinition of the Flank Zone.

Redefining the flank

The treaty loophole that made redefining the Flank feasible was that Article V defines the Flank Zone in terms of Soviet Military Districts as they existed in November 1990. The treaty does not further define the Military Districts which in any case were no longer Soviet, but a mixture of Russian and Ukrainian. Without changing the language of the treaty, it would thus have been feasible to redefine the Military Districts to meet Russian and Ukrainian concerns about Flank sub-limits.

While not explicitly redefining their MDs, in December 1994 at the annual exchange of CFE data, Russia warned that it would not in future be exchanging separate information on equipment in the southern part of the North Caucasus Military District. Material in this 'restricted zone of operations' would be included in Russia's overall national ceilings, but not in the Flank sublimits. The zone which Russia effectively wants removed from Flank limits stretches from the Russian border with Azerbaijan and Georgia, north up the Black and Caspian Sea coast, ending in a line coincident with latitude 45 degrees 15 minutes; an area which includes Chechnya.

Over the summer of 1995, NATO's High Level Task Force developed a proposal to redefine the Flank which it presented to Vitaly Churkin, Russian ambassador to NATO in September. NATO sought to meet the Russian concerns by removing five Oblasts from the Flank to the Rear Zone: to (Volgograd and Astrakhan) from the North Caucusus MD; and three (Pskov, Novgorod and Vologda) from the Leningrad MD. In a similar gesture to Ukraine, NATO proposed to remove the Oblast of Odessa from the Odessa MD.

NATO indicated that the *quid pro quo* from Russia and Ukraine should include a willingness:

- to accept more inspections
- to provide more information than currently required under CFE
- to exercise restraint in force deployment, for example, no destabilising concentrations near borders with other CFE State parties
- to reaffirm the Soviet unilateral statements of June 1991 with respect to restraint on coastal defence forces and naval infantry[25]

Russia responded with a counter-proposal in October which was unacceptable to NATO. There was thus no agreement on redefining the Flank before the reduction period ended on 17 November. This left Russia and Ukraine in technical violation of the treaty in the run-up to the CFE Review Conference scheduled for May 1996.

Issues for the review conference in 1996

Assuming optimistically that CFE survives the Slavic slippage, a number of other issues will be raised at the Review Conference scheduled for May 1996. These include whether to revise or maintain the same group and national ceilings as NATO takes on new members; whether to exempt from CFE limits equipment used in peacekeeping operations in the CFE zone of application, as Russia suggests; and whether and how to harmonise CFE limits throughout the CSCE region as France wants.

Maintaining the same NATO group limits as new members join

the alliance would be one way to demonstrate to Russia that an enlarged NATO poses no greater military threat than the NATO of the sixteen. Western governments were wary of this idea in mid-1995, but many western officials also acknowledged that CFE limits were far too high, leaving plenty of headroom to accommodate new members under the old group ceiling.

Exempting from CFE limits equipment used in peacekeeping operations is also an idea worth pursuing, although parties to CFE would be wary of using Russian operations in the Caucasus as a precedent. Only Russia defines that operation as peacekeeping; the international community saw it as a brutal civil war. If, however, CFE parties participate in genuine CSCE or UN-endorsed peacekeeping operations in the zone of application, then exemptions seem at least worth considering.

Harmonising CFE limits throughout the CSCE area will be much more difficult to manage, but any settlement in the Balkans will only be credible if it imposes a stringent arms control regime on Serbia and Croatia. Binding all the former Yugoslav states into CFE would be one way to do this. In which case the scope of limitations would need to include the lower gauge artillery and lighter equipment with which the Balkan war was waged.

Means will be as important as ends in the Review Conference negotiations. If, as seems likely, it will be held parallel to discussions, if not implementation, of NATO enlargement, CFE revisions should manifestly address Russian concerns about new NATO members, while not undermining the security interests of either NATO or Russia's smaller more vulnerable neighbours. CFE revisions should be confidence and security building measures undertaken in the spirit of conciliation and cooperation in which CFE was negotiated in the first place.

Britain played a leading role in the original CFE negotiations, developing especially good relations with the central European countries. It will need to play as vigorous a role in this next round of talks because the stakes are so much higher and there is little leadership on this issue from Washington. Cooperation with Germany will be vital, especially in bringing France to accept a common NATO position.

14 PRIORITIES FOR NUCLEAR, CHEMICAL AND BIOLOGICAL ARMS CONTROL

Stephen Pullinger

Britain faces an increasingly complex international environment in which new non-military, as well as more traditional military, threats and challenges can be expected to emerge. One threat that is already evident is the proliferation of weapons of mass destruction. The challenge for Britain, as an acknowledged nuclear weapon state (NWS) and leading proponent of chemical and biological disarmament, is to pursue a coherent arms control strategy that assuages that threat.

Britain possesses neither chemical nor biological weapons and has traditionally played a leading role in international efforts to abolish them worldwide. On the other hand, the British government insists on its need to retain and modernise its nuclear weapons and cannot foresee a time when these might be relinquished.

This chapter seeks to demonstrate that Britain can play a positive role in the nuclear disarmament process, in much the same way as it has in chemical and biological weapons disarmament. The practical measures recommended would uphold Britain's defence interests in the near term whilst contributing to its enhanced security in the long term.

Chemical weapons disarmament

There is a sharp contrast between Britain's nuclear weapons and nuclear arms control policies and those it pursues with regard to chemical weapons and chemical arms control. Britain unilaterally abandoned its chemical weapon capability in the late 1950s and has long been a major advocate of chemical arms control. Following a lengthy negotiation, in which Britain played a leading role, the Chemical Weapons Convention (CWC) was signed in January 1993. The CWC outlaws the development, production, stockpiling and use of chemical weapons. Now signed by over 150 states the CWC represents the most comprehensive and intrusive arms control treaty so far agreed.

The Convention requires the declaration and destruction of existing chemical weapons stockpiles and production facilities. It also seeks to prevent states from resuming or starting a chemical weapons programme through the application of an extensive verification regime, controlled by the Organisation for the Prohibition of Chemical Weapons (OPCW). This body is vested with extensive legislative, law enforcement and judicial responsibilities. On-site monitoring of certain chemical facilities are provided through permanent installation and inspection. Inspections can be conducted on a challenge as well as a routine basis.

Each state party to the CWC is responsible for ensuring that it fulfils its obligations under the treaty. Therefore, Britain will have to establish its own national authority to act as a focal point for liaison with the OPCW and other states parties. Its duty to ensure that the UK remains within the terms of the treaty could be an extensive one, especially if it has to exercise general surveillance over *all* toxic chemicals and their precursors, of which many millions are currently known.[1]

The CWC will only enter into force 180 days after the 65th instrument of ratification has been deposited. By November 1995, only 27 state signatories have done so, including France and Germany. Those that have yet to ratify include the US, Russia and the UK. Republican gains in the US Senate have slowed the American ratification process, whilst Russia is reluctant to commit the vast sums needed to implement the treaty that its ratification would require. Britain's tardiness in ratifying is less easy to explain, given the high priority chemical arms control has traditionally enjoyed. Before ratification is possible, Britain needs to introduce enabling legislation to allow, amongst other things, international inspectors rights of access to commercial chemical facilities to ensure verification of compliance. The Department of Trade and Industry, rather than the Foreign Office, is responsible for initiating this legislation and for the subsequent creation of the national authority.

According to the 1995 Defence White Paper, around a dozen countries of concern either have or are developing some form of weapon of mass destruction. The most notable of these are in the Middle East, where a number of states have refused to sign the CWC, justifying their decisions largely as a response to Israel's nuclear capability and that country's refusal to accede to the NPT.

It is in Britain's interest that the CWC should enter into force at the earliest opportunity, for it to be properly enforced and for those states that have yet to sign to be persuaded to do so. To assist that process it is important that Britain become an original state party to the treaty and then plays a leading role in its implementation.

Biological weapons disarmament

The Biological and Toxin Weapons Convention (1972) bans the development, production, stockpiling, acquisition or retention of lethal biological agents, as well as the development or transfer of biological munitions or delivery systems. Unfortunately, the Convention lacks a verification regime. Each subsequent review conference, held every five years, has recognised the need to address this problem.

In late 1995 there were 133 State Parties to the Convention and a further 18 signatories. However, there are concerns that about 10 countries may be pursuing actual biological weapons (BW) programmes. As biological agents are much more potent than chemical warfare agents these capabilities represent a major potential threat to international security. Indeed, biological weapons are becoming increasingly usable instruments of warfare. Nor is there any longer any real necessity to have a stockpile of actual weapons, as the BW agent can be produced in sufficient quantity readily and rapidly and then dispersed using a simple spray system.[2] This merely adds to the problem of verification.

In September 1994 a Special Conference was held in Geneva to consider establishing a verification regime, and an *ad hoc* group was created to take this work forward. The group was mandated to draft proposals for strengthening the Convention with a view to their inclusion in a legally binding instrument.[3] Britain has already conducted a series of practice inspections in the biotechnology and pharmaceutical industries to examine verification procedures and to explore the implications for industry of intrusive inspections.

Countering the proliferation of biological weapons is a matter of extreme importance and strengthening the BW Convention through an effective verification regime, and persuading more states to join, offer the best hope of success.

Nuclear arms control

Which way forward?

It is possible to identify three broad approaches to British nuclear arms control policy:

(i) Unilateral disarmament
A sizeable minority of British public and political opinion opposes possession of nuclear weapons. That opposition is based essentially on two principles. The first is that since the use of nuclear weapons would inevitably lead to massive civilian casualties such an act could never be justified and should never be threatened. The second is that deterrence is a flawed concept which fosters the adoption of an irrational policy of threatening to commit national suicide. Therefore, Britain's nuclear weapons should be abandoned regardless of what other states choose to do.

(ii) The British Bomb for ever
The 1990s Conservative government places Britain's nuclear weapon programme in a category all of its own; something to be supported uncritically, and developed in isolation from any other policy considerations. Important as nonproliferation policy is, for instance, it is not allowed to impinge upon the requirements of the national nuclear weapon policy. Consequently, arms control priorities are driven by the weapons programme rather than by wider concerns.

Hence, the British government initially fiercely resisted an end to nuclear testing because the needs of its weapons programme took precedence over any nonproliferation benefits accruing from a Comprehensive Test Ban Treaty.

(iii) Active Multilateralism
To date, these two polarised views have dominated the nuclear weapons debate in Britain. The first sees nuclear weapons as immoral, useless objects with no military or deterrent attributes, which should be scrapped immediately regardless of arms control negotiations. The second regards them as weapons imbued with almost mystic war-preventing qualities, that should be endorsed as essential prerequisites of national security. This unrealistic dichotomy prevents pursuit of a multilateral arms

control strategy towards the establishment of an international security regime in which no country feels threatened by nuclear weapons, no country sees it as being in its interest to acquire them and the only purpose of such weapons, if any, is to deter other nuclear weapon states.

These security criteria should guide arms control policy in the short- to medium-term rather than a predetermined commitment either to retain nuclear weapons indefinitely or to achieve their complete abolition. What is clear today is that we can move a lot further down the path of denuclearisation before we need to decide whether or not we can safely make that final leap to complete abolition. More importantly, we can enhance national and global security in so doing. Supporters of this approach, therefore, should include those that believe in a nuclear-weapon-free world as well as those that remain highly sceptical.

Recent British governments were too dismissive of the links between their nuclear weapons policies and the health of the nonproliferation regime. It is counter-productive in the struggle against proliferation for example, when Britain demands more intrusive inspections of non-nuclear weapon states whilst refusing to divulge exact information concerning its own weapon deployment; or to call for indefinite nuclear weapon abstinence by others whilst displaying an obvious reluctance to contemplate a time when Britain might become a non-nuclear weapon state.

By being an active and enthusiastic participant in the disarmament process, however, the next British government could achieve a great deal in nonproliferation terms. Unilateralist sceptics of this approach should appreciate that whereas multilateralism may have been used as an excuse for disarmament deadlock in the past it now provides many opportunities for significant progress. Britain could undertake a range of arms control, confidence-building and disarmament measures to enhance both its own and international security, and command widespread popular support.

We have reached the point where certain traditional aspects of nuclear weapons policy and doctrine may now have detrimental consequences for this country's long-term security by obstructing arms control initiatives and thereby undermining the imperatives of our own nonproliferation policy. If the British government is to adapt to the new approach suggested earlier, these policies will need to be harmonised.

A new nuclear arms control, confidence-building and disarmament strategy

The Conventions concerning chemical and biological weapons are already in place and, as stated above, need to be strengthened and universally adhered to. Britain's arms control record in both these areas is probably second to none. It is in the nuclear sphere that a new arms control strategy is required.

If one accepts the premise that British nuclear weapons perform a limited function in the last resort it is possible to examine what further measures of arms control, confidence-building and disarmament Britain could safely undertake. To reiterate, the purpose of that programme of steps should be to support the establishment of an international security regime in which no country either feels threatened by nuclear weapons or sees it as being in its interests to acquire them, and the only purpose of nuclear weapons, if any, would be the deterrence of other nuclear weapon states.

The major legal requirement pertinent to Britain's arms control policies is provided by the obligation under Article VI of the NPT to:

> pursue negotiations in good faith on effective measures relating to cessation of the nuclear arms race at an early date and to nuclear disarmament, and on a treaty on general and complete disarmament under strict and effective international control.

John Major's government says that it is 'committed to work towards nuclear disarmament'[4] and yet, when asked why HMG would not put all British nuclear weapons into disarmament negotiations, the then Defence Secretary Malcolm Rifkind said: 'First, that is not a commitment under the non-proliferation treaty and, secondly, it would be a remarkably foolish initiative...'[5]

This is a government that acknowledges its treaty obligations, but is reluctant to carry them out. But a new British government could actively work towards nuclear disarmament in a number of ways.

Security assurances

The purpose of security assurances is to reassure those who do not have nuclear weapons that they will never be subject to nuclear threat or attack and that if they are, the other NWS would come to their assistance. These assurances should thereby reduce the incentive to states to acquire nuclear weapons of their own.

A collective positive security assurance was provided by the US, UK and Russia in 1968 under UN Security Council Resolution No. 255. This was updated in April 1995, when China and France also agreed to a common text. This pledged that any state which commits aggression accompanied by the use of nuclear weapons or threatens such aggression will be subject to measures taken in accordance with the UN Charter to 'suppress the aggression or remove the threat of aggression'.

Britain reaffirmed its intention to seek immediate UN Security Council action to provide assistance to the victim of such aggression. This assistance could include measures to 'settle the dispute and restore international peace and security'. Britain would also be prepared to offer the victim of nuclear attack 'technical, medical, scientific or humanitarian assistance'. Precisely what measures might be involved in restoring peace would obviously depend on the particular circumstances. Non-NWS Parties to the NPT pressed the NWS to give more concrete assurance that they would act decisively against any future nuclear protagonist who acted in this way. However, given political realities, it is difficult to see how these assurances could go much further without beginning to lose credibility.

There is also an extremely important limitation on the possible use of British nuclear weapons. It is contained in the UK's negative security assurance (NSA), originally provided in 1978, and updated and re-issued in April 1995 as what is essentially a common text with the US, Russia and France. It states that the UK will not use its nuclear weapons against any non-nuclear weapon state parties to the NPT except in the case of an invasion or an attack on the UK, its dependent territories, its armed forces or other troops, its allies or a state towards which it has a security commitment, carried out or sustained by such a non-nuclear weapon state in association or alliance with a nuclear weapon state. China issued a separate, stronger assurance that pledged itself never to use nuclear weapons first 'at any time or in any circumstances'.

The purpose of Britain's assurance is to allay any fears of non-nuclear weapon states that British nuclear weapons might be used against them, and to restrict their utility purely to the deterrence of other nuclear weapons. For example, the use of British nuclear weapons against Argentina during the Falklands war, or even against Iraq during the war over Kuwait, would have been 'inconceivable', according to the government.[6] The exception attached to Britain's NSA was originally intended to cater for aggression by non-nuclear weapon state members of the Warsaw Treaty Organisation (WTO), acting in concert with the nuclear-armed Soviet Union.

As with positive assurances, many non-nuclear weapons states parties to the NPT feel that the negative security assurances provided by the nuclear weapons states (with the exception of China) is inadequate. There are two obvious ways in which it could be strengthened. The first would be for the US, UK, France and Russia to drop their exemptions and make unequivocal NSA declarations, whereby they would never use their nuclear weapons against a state that did not possess such weapons itself. Britain could agree to do so on the grounds that there would appear to be no conceivable circumstances *at present* in which the existing exemption would apply. The option to reverse that concession should the security situation deteriorate sufficiently to warrant it, would always remain available.

However, the present British government has explicitly rejected calls to remove the qualifications attached to its NSA, insisting that they should remain in order to deal with any future replication of the WTO: 'We would have to provide against the possibility that a non-nuclear weapon state would get into bed with a nuclear state and threaten us'.[7]

The other way to strengthen security assurances would be for Britain to commit itself never to use its nuclear weapons first. By definition, if everyone adopted a No First Use (NFU) policy and adhered to it the use of nuclear weapons would be rendered impossible. Indeed, this is one of the main motivations of those that seek a common NFU policy. It also offers another means of obtaining assurances which are perceived to be lacking in NSAs.

The other rationale behind an NFU policy arose from concerns that the superpowers were striving to attain a first strike capability, whereby one side could destroy sufficient of the other side's nuclear forces to render the subsequent retaliatory blow 'insignificant' or 'acceptable'. Nowadays, with the Cold War over, such fanciful ideas have been shelved. Moreover, after START II the possibility of launching a successful first strike will become even more unlikely (with the removal of all MIRVed ICBMs). So in this sense an NFU policy is no longer so relevant. This is certainly true for Britain, because it neither possesses nor aspires to possess a first strike capability.

Opponents of an NFU policy argue that it offers unnecessary assurances to potential nuclear-armed aggressors by telling them that even if they attack Britain with conventional forces, and even if they are winning the war or achieving their military objectives, Britain will not use nuclear weapons against them. In doctrinal terms, therefore, an NFU policy almost wholly contradicts the basis of sub-strategic use, the purpose of which is to signal intent to an aggressor through the initial and limited use of nuclear weapons.

In reality, the only nuclear-armed adversary with considerable conventional forces at present is Russia (China has substantial forces, but is too far away to be of direct concern to Britain). Yet, since the collapse of communism, the disbanding of the WTO, the breakup of the USSR and the signing of the Conventional Forces in Europe (CFE) treaty there is now a conventional imbalance of forces in favour of NATO and the central European states against Russia.

This present disposition of forces thus opens the way for NATO to reconsider an NFU policy because it could expect to win any conventional war in Europe without resorting to the use of nuclear weapons. Of course, if at some future date Russia reconstituted its forces to the point where invasion of western Europe was a viable option once more (it would have to break its CFE treaty obligations to build up such a capability), NATO would feel entitled to consider withdrawing any NFU commitment it might have made.

Under present circumstances a reciprocal NATO–Russia NFU agreement would be a positive further step towards building security between the two, in a similar way to the Anglo-Russian nuclear de-targeting accord. Britain can work towards securing such an agreement (conditional on the maintenance of the improved security situation in Europe) and be prepared to pursue a strengthening of the existing common NSA.

The UK government's resistance to changing its attitude was enunciated by the then Foreign Secretary, Douglas Hurd, in testimony to the Commons Foreign Affairs Select Committee in January 1995, when asked if the government proposed to give up the option of first use, answered:

> No, we have to maintain it. We have to maintain that option not because we see those circumstances of a huge conventional threat repeating themselves quickly, but one can never be sure, and this whole business is to deal with the uncertainties of the international arena.[8]

Reassessing a minimum deterrent

Before entering any arms control process, Britain should re-assess exactly what level of nuclear capability it needs in order to maintain a credible minimum nuclear deterrent. (This issue is discussed by Michael Clarke in Chapter 16.) Britain should beware of giving the impression that it is exploiting its privileged position as an acknowledged nuclear weapon state. Nor should it act in a manner contrary to the spirit, if not the strict letter, of Article VI of the NPT.

Trident

The replacement of the UK's Polaris fleet of four ballistic missile-carrying submarines (SSBNs) with four Trident SSBNs is already underway. It provides scope for a massive increase in nuclear capability. Each Polaris boat carries no more than 48 warheads on its 16 missiles, capable of hitting one target per missile, at a maximum range of 4,600 km. On the other hand, each Trident boat's 16 missiles can deploy a total of up to 192 warheads, each of which is capable of hitting a separate target up to 12,000 km away, and with far greater accuracy than Polaris.

The level of capability deployed (in terms of numbers of missiles and warheads), however, is entirely a matter of military and political judgement and decision. Each Trident D5 missile can carry any number of warheads between one and 12. In 1995, the Conservative government announced its decision to deploy a maximum of 96 warheads on each Trident submarine, but refused to divulge the exact number actually deployed. It has, however, stated that each boat will deploy with an explosive power 'not greatly in excess' of that deployed on Polaris and said that Trident will deploy with 'fewer than 300 operational or available warheads'. It claims that this level of capability constitutes a minimum deterrent. It has also revealed that the Trident force will comprise 30 per cent fewer operational or available warheads than the total number deployed on Polaris and with the RAF during the 1970s. As only three submarines are loaded at any one time, this allows for a maximum of 288 warheads. However, the actual figure may well be closer to 200.[9]

The Labour and Liberal Democrat parties have challenged this doubling of the UK's SSBN warhead ceiling, arguing that if no more than 48 warheads per boat was an adequate deterrent during the Cold War, why are up to twice that many needed now the Cold war is over? They believe that the number of warheads on the Trident system should be no higher than that deployed on Polaris.

There is no indication that the Ministry of Defence's plans for the level of the Trident deployment have altered since the programme was first envisaged. It is quite conceivable that the MoD has always been planning to deploy an equivalent 'explosive power' on Trident to that on Polaris. An announcement in November 1993 that the originally stated maximum warhead deployment on each Trident boat would be 96, rather than the previously stated 128, was interpreted by many as a reduction in the planned actual deployment. Yet this does not necessarily follow. When asked if the change in the ceiling figure constituted a change in the actual number of warheads or a clarification (of what had previously been decided), the Director of Nuclear Policy and security, MoD, Mr Thatcher, replied 'It is a clarification'.[10] As this formula still allows Britain to deploy a system with far greater targeting flexibility, improved accuracy and more, smaller yield, warheads, the Conservative-dominated Defence Select Committee was led to conclude that the planned Trident deployment 'does—and was always intended to—represent a significant enhancement of the UK's nuclear capability'.[11]

It is time for the level of Britain's minimum nuclear deterrent to be re-examined, in light of the enormous changes in the threat environment since Britain's Trident programme was first conceived. Until a few years ago government ministers were willing to justify the planned warhead increase on the basis of needing to maintain the ability to defeat anti-ballistic missile (ABM) defences. Since the only ABMs in existence are those around Moscow, this was an implicit signal that Russia was the crucial target for British nuclear weapons. Yet relations with Russia have since improved to the point at which the British government has felt able to sign a bilateral agreement with Russia not to target each other's territory with nuclear weapons. Although the government no longer publicly speaks of the need for Trident to penetrate ABM defences, it is quite likely that satisfying the 'Moscow Criterion' remains a key requirement of this government's nuclear doctrine.

When Moscow was the heart of a centralised Communist state the ability to destroy it may well have been a crucial component of deterring the USSR. In the new circumstances a minimum deterrent could probably comprise a capability to destroy, say, a dozen Russian cities other than Moscow (which are undefended by ABMs). In terms of the *number* of targets that can be hit, the government's plans allow for up to a sixfold increase in capability. On the face of it, therefore, it would appear that there is scope for further restrictions to be imposed on the Trident warhead deployment.

The maintenance of any new, lower warhead ceiling for the Trident boats would depend on continued adherence to other arms control agreements, most especially the ABM treaty. In order to be able to respond to any serious deterioration in the nuclear arms control regime, such as a break-out from ABM treaty restraints, which could undermine the credibility of Britain's small nuclear force, a British government would be wise to retain the capacity and means to expand its operational warhead deployment.

Although the number of warheads is the most relevant measure of nuclear capability, other aspects of the Trident programme can be re-examined to see if further cost savings could safely be achieved. For instance, under present plans, the present Conservative government is planning to purchase an additional 21 D5 missiles (between FY 97-99) from the US to complement the 44 already bought. If Britain were to restrict further its warhead deployment, then it may well be able to make do with fewer than the 65 missiles it currently requires. Instead of loading 16 missiles per submarine, 12 might suffice. This could save at least £300 million pounds on the missile purchase.

A reduction in the number of crews that support the SSBN fleet might be possible (an option already under consideration by MoD).[12] The availability of much greater warning time of attack could allow one of the four submarines to be kept in a state of extended readiness, rather than as part of the operational cycle.

This extra warning time has led some to suggest that it is no longer essential to maintain at least one submarine permanently on station and that only three submarines are really necessary. The MOD counters by arguing that if one of the boats is always in extended refit, a three-boat fleet offers no margin for accident or serious failure in one of the others. In those circumstances, Britain might well not be able to maintain one boat on operational patrol. Proponents of three boats are prepared to accept the risk of this eventuality, but the MoD wants the extra insurance which the fourth boat provides.

There is a risk that lowering that standard of preparedness may send the wrong political signals that Britain is only going through the motions of maintaining a nuclear deterrent. This, in turn, could feed through into the attitudes and day-to-day work practices of those who operate and maintain the fleet. It is for these reasons, rather than any expectation of a bolt-from-the-blue attack (although this can never be entirely ruled out), that it is probably unwise to relax this accepted deployment benchmark. The cost savings would probably not offset the perceived devaluation in deterrent terms.

WE-177

Britain's sub-strategic nuclear capability currently consists of an unknown number (probably about 50-75) of ageing WE-177 nuclear free-fall bombs. Having decided against a tactical air-to-surface missile (TASM), the present government said that when the WE-177 leaves service, Trident will take on the sub-strategic role. In April 1995 the government announced that the remaining WE-177s would be withdrawn from service by the end of 1998.[13] This is to coincide with the entry into service of the third Trident submarine, although the MoD admits that Trident could provide a continuously available sub-strategic capability when the second submarine joins the patrol cycle around the end of 1995. The reason why the WE-177 will stay until the third boat is operational is to insure against an accident or serious fault occurring to one of the other two.

Transparency

It is time for Britain to state explicitly the number of nuclear warheads it deploys operationally. Its current refusal to do so is a legacy of old Cold War suspicions that disregard the value of such a declaration as an important confidence-building measure in the context of the NPT. The NWS are right to press for a greater degree of openness and access to the nuclear activities and facilities of non-nuclear weapon states to ensure that they are not developing nuclear weapons. But in the same spirit, they should be open about the number of weapons they deploy.

The government's decision to reveal that when Trident is fully deployed Britain will have fewer than 300 operational or available warheads is a welcome step in the right direction, but still falls short of providing a full disclosure of actual warhead numbers deployed. To this end, the German initiative to establish a nuclear arms register should be developed and all five nuclear weapon states invited to participate. The number of warheads declared by Britain could become a conditional maximum ceiling, and act as the benchmark measurement for involvement in actual disarmament negotiations.

Britain should undertake not to exceed its ceiling unless other states attempted to deploy the means to challenge the deterrent effect of that level of capability.

Strategic Arms Reduction Talks (START)

If and when START II is successfully ratified by the US and Russia the way will be open to consider the negotiation of a START III treaty. Unlike the previous START agreements, START III is likely to take account of, if not actually include, the nuclear weapons of Britain, France and China as well as those of the US and Russia. The Clinton administration has been examining START III options and in February 1995 the Russian Ambassador to the Conference on Disarmament reiterated Yeltsin's call at the UN for a treaty on nuclear security and strategic stability. The Ambassador proposed a phased approach whereby a variance of obligations would be expected of states with different sized arsenals:

> During the first stages, while Russia and the US would agree on further steps to reduce their nuclear forces, other nuclear states could take the commitment not to increase their existing nuclear armaments.[14]

However, at the NPT Extension Conference the British government set out the necessary conditions for its involvement in strategic disarmament negotiations:

> a world in which the nuclear forces of the Russian Federation and the USA were numbered in hundreds rather than thousands would be one in which the UK would respond to the challenge of multilateral talks on the global reduction of nuclear arms.[15]

Because the next START agreement is unlikely to reduce US and Russian arsenals below 2,000 warheads each, this strongly suggests that the present British government wants to exclude itself from participation in the next round of disarmament talks. As the Russian Ambassador outlined, because of the disparity in the relative size of the US and Russian arsenals compared to those of the UK, France and China, START III can be expected, initially at least, to concentrate on further reductions in US and Russian arsenals, whilst imposing some type of 'cap' on the arsenals of the other three. If this approach were adopted it need not be incompatible with Britain's present position.

Britain, along with the other NWS, firmly resisted the acceptance of any time-frame being imposed on their pursuit of disarmament, although they did all agree a common set of principles at the NPT Conference. This included an agreement to 'the determined pursuit...of systematic and progressive efforts to reduce nuclear weapons globally, with the ultimate goal of eliminating those weapons'. This language was sufficiently open to interpretation for John Major's administration in Britain to justify its cautious approach to nuclear disarmament. However, the text could also be enthusiastically embraced by a subsequent British government that was more committed to nuclear disarmament and used to spur more rapid progress towards that end.

Obviously, if a British government had already implemented a further curtailment of Trident and abandoned the WE-177 it would have reduced the UK's forces to a level below which it could not fall without beginning to undermine the credibility of its deterrent. There would be very little scope, if any, for further reductions until one reached the stage (if one ever does) where complete nuclear disarmament becomes a serious and acceptable proposition. However, subjecting those new restrictions to treaty controls would allow them to be properly verified, would embed them in a legal treaty regime and would, therefore, represent a significant step forward.

Comprehensive Test Ban Treaty (CTBT)

A CTBT would be a major boost to the nonproliferation regime for two main reasons. First, because it would demonstrate to the non-NWS that the NWS are serious about their obligations under Article VI. Second, because although it would not stop the development of first generation atomic bombs, it would severely hamper the successful development of second generation weapons and the miniaturisation of devices necessary for deployment on ballistic missiles, for example.

Yet the present British government initially strongly resisted renewed efforts to achieve a comprehensive ban on nuclear testing. It did so because it still saw a CTBT in terms of its impact on Britain's nuclear weapon programme rather than its nonproliferation benefits (despite the fact that the testing of Trident warheads had already been completed). Only when it became clear that the US was not going to allow further British tests at Nevada did the government reluctantly accept that if Britain could not test again then it would be better if no-one else could either.

This was a clear example of where old-style thinking obscured the nonproliferation priorities of the new international reality. Britain's unhelpful insistence on retaining a clause in the draft treaty text that would allow tests in exceptional circumstances even under a CTBT was dropped in April 1995. In September 1995, Britain accepted a zero-yield test ban and, along with the other nuclear weapons states, is now committed to the conclusion of a CTBT by the end of 1996.

Fissile material cut-off

A cut-off in the further production of fissile material for weapons purposes is another proposal with beneficial nonproliferation implications, which should have no detrimental impact on the maintenance of Britain's minimum nuclear force.

The government had already declared its readiness to participate in negotiations on a multilateral convention, when it subsequently also announced that Britain had stopped the production of fissile material for weapons purposes at the NPT Conference in April 1995. This became possible because Britain already has sufficient fissile material to satisfy its existing weapon requirements as well as any foreseeable operational needs. Ministers have also made clear that Britain will continue to recycle such material from dismantled weapons into new Trident warheads. A call by Germany, amongst others, at the NPT Conference to forbid recycling was firmly rejected.

The way was cleared for talks on a fissile cut-off to proceed following agreement on a negotiating mandate in early 1995. However, existing military stockpiles of fissile materials are excluded from the mandate, which devalues the importance of any eventual agreement. As with its initial reaction to the new push for a CTBT, Britain's opposition to widening the scope of the convention is indicative of nonproliferation imperatives taking a back seat to the demands of the national nuclear weapon programme. Ultimately, unless Britain pursues a positive multilateral approach to nuclear disarmament, it may undermine the nonproliferation regime upon which its future security depends.

15 CONFIDENCE BUILDING THROUGH VERIFICATION AND TRANSPARENCY

Patricia M. Lewis

This chapter outlines the principles of verification, looks at its history and details a number of proposals which the next British government could put into practice for very little cost whilst greatly increasing international security.

As a technologically advanced country with a permanent seat on the United Nations' Security Council Britain is in a position to play a leading role in the verification of arms control agreements. To date, however, British governments have been reluctant to open up nuclear weapons facilities and deployment platforms to international inspection. Nor has Britain yet ratified the Chemical Weapons Convention, apparently because of reluctance to allow inspection of chemical industries.

This is a short-sighted policy since verification is the process which establishes whether all parties to an agreement are in compliance. Reliable verification measures are essential to generate the atmosphere of trust on which international arms control regimes depend, because states are more likely to implement agreements fully when they are confident that cheating by any other party is likely to be detected. Verification is thus beneficial to all parties and to both national and international security interests. Defence funds invested in verification techniques are always positive sum investments. In this they differ from spending on new weapons systems which may be justified to enhance British security, narrowly defined, but often have a negative sum outcome by triggering security anxieties in neighbouring states.

In addition to investing more research and development funds in verification measures for arms control Britain could also pioneer the application of arms control verification techniques to conflict prevention, especially to the building of confidence between potential adversarial groups among the newly emerging former Communist states. There are also many regions of the world (Europe, South Asia, the Middle East, the North Pacific and Southeast Asia) where Britain still has influence and could promote regional confidence-building and security-building measures.

What is verification?

Verification is a process which establishes whether all parties to an agreement are complying with their obligations. These agreements can be international treaties on arms control or the environment, or agreements between different communities within a state. The success of any agreement depends on building an atmosphere of trust which can be built and maintained when all sides are aware that cheating is likely to be detected.

It must be noted, however, that there is no such thing as 100 per cent certainty in verification. Verification measures are designed to ensure that a party contemplating cheating on a treaty cannot do so without running a substantial risk of being found out. Regimes can be designed so that the likelihood of catching significant cheating is very high (say, 80–100 per cent) or is low (say, below 50 per cent). Generally, the more effort, money and resources put into verification, the higher the probability of detecting cheating.

The process includes the collection of information relevant to obligations under arms limitation and disarmament agreements; analysis of the information; and reaching a judgement as to whether the specific terms of an agreement are being met.[1] Verification *per se* is treaty/agreement specific. Monitoring—the open collection of information—can be part of the verification process or it can be quite separate. (for example, the 1992 multilateral Open Skies agreement); and can be separate from any agreement (for example, intelligence gathering). However, the difference between verification and monitoring is becoming increasingly obscure. The purpose of verification is to make it unacceptably risky for any party to cheat on an agreement. If the verification provisions of an agreement are comprehensive, then parties will be deterred from cheating because they know that they run a high risk of getting caught. This is called 'Verification Deterrence'. In this way agreements can be 'built to last' and states or communities can build other elements of their security framework based, in part, on the knowledge obtained through comprehensive verification.

Intelligence and verification

There is a synergy between verification of arms limitation agreements and intelligence gathering for national security. Both processes include collecting information, collating information from a number of sources, analysing the information and distributing the information or analysis to interested parties.[2] Both verification and intelligence activities lead eventually to decisions on national and international security. The key difference between verification and intelligence gathering is that the former is carried out entirely in the open with the consent of all participating states whereas the latter is a highly secretive operation. However, intelligence agencies play a role in verification, often by providing background information or by making suggestions for on-site inspection targets. The verification process also feeds information into the intelligence agencies such as 'ground-truthing' (establishing if the information on the ground supports the information gleaned from satellites).

Secret intelligence gathering can be useful for the purposes of verification, and information gleaned through verification is always useful for the intelligence agencies. However, secret monitoring can also be a liability for verifiers because, if there is suspicion that, say, an on-site inspection is being carried out for reasons not to do with the treaty, or if one of the inspectors is clearly an intelligence agent and acting for the agency, then the whole process of verification could be brought into disrepute and states would no longer participate in a cooperative fashion.

Recent verification history

USA–USSR (Russia) bilateral treaties

From the end of the Second World War and the beginning of nuclear weapons until the rise of Mikhail Gorbachev in the USSR, the role of verification in US–USSR arms control treaties was greatly dependent on the technology available to carry out monitoring at a distance. Throughout the bilateral negotiations, up until the 1987 Intermediate-range Nuclear Forces (INF) Treaty, the issue of intrusive on-site inspections for verification purposes was guaranteed to stall or even halt negotiations. The USA pursued the concept of 'anytime, anywhere' inspections,[3] whilst the USSR viewed such proposals with intense suspicion, believing inspections to be cover for espionage. The stand-off was so established that a US arms control expert claimed that 'verification is becoming a shield for those not interested in arms control to hide behind'.[4]

The Strategic Arms Limitation Treaties and the Anti-Ballistic Missile Treaty all relied for their verification on 'National Technical Means' which, in the arms control context, meant monitoring by intelligence satellites. The Threshold Test Ban Treaty (TTBT) and the Peaceful Nuclear Explosions Treaty (PNET) were not ratified for many years partly due to the issue of verification. Not until 1991, when agreement was reached on intrusive verification, did these treaties enter into force.

The breakthrough in intrusive verification between the two superpowers came when Gorbachev introduced the policy of '*glasnost*' (openness) and offered to open up sensitive military sites for inspections. The first bilateral agreement which took advantage of this change in policy was the 1987 INF Treaty, which not only included on-site inspections to INF bases but also allowed monitoring of production facilities and of missile reduction. Since then the USA has backtracked,[5] and the military and commercial agencies have worried also about the intrusiveness of on-site inspections and the cost of the verification regimes.

Multilateral agreements

The main multilateral agreements in the field of arms control and disarmament are as follows: Geneva Protocol (1925, entry into force–eif–1928); Antarctic Treaty (1959, eif 1961); Partial Test Ban Treaty (1963); The Outer Space Treaty (1967); Tlatelolco (1967, eif 1968); Non-Proliferation Treaty (1968, eif 1970); Seabed Treaty (1971, eif 1972); Biological Weapons Convention (1972, eif 1975); Enmod Convention (1977, eif 1978); Inhumane Weapons Convention (1981, eif 1983); Rarotonga (1985, eif 1986); Stockholm Accord (1986); Conventional Forces in Europe (1990, eif 1992); Vienna Document (1990); Open Skies (1992 yet to enter into force); Chemical Weapons Convention (1992 yet to enter into force).

Although East–West tensions were played out in multilateral negotiations (for example, in the Chemical Weapons Convention [CWC], their effects were often mitigated by states not participating in the Cold War. As a result, the arguments over intrusive verification, particularly on-site inspection, were of a different calibre.

Treaties such as the 1968 Non-Proliferation Treaty (NPT) and the 1959 Antarctic Treaty have provisions for on-site inspection although the 1963 Partial Test Ban Treaty (PTBT) and the 1972 Biological Weapons Convention (BWC) have no verification provisions at all.

In 1986, before the end of the Cold War but during the Gorbachev thaw, the highly significant Stockholm Accord was agreed between the participating states of the CSCE. This was a series of confidence-building measures designed to increase transparency over military exercises in Europe. From the beginning, states carried out challenge inspections of military exercises—information on the exercise calendars being exchanged between the states. The execution of the Accord was very successful. The trust which built up between the CSCE countries as a result of the Stockholm Accord had a number of effects including: (i) the formation of friendly relationships between east and west inspectors; (ii) a shift in perception of 'the other side' as 'enemy' so that there was a sense of common purpose; (iii) a pride in the inspection process itself—this led to friendly rivalry in, for example, seeing which team could offer the best food and wine etc.; (iv) a reduction in the number and scale of the military exercises (partly as a result of lessening tension and partly as a result of cutting the cost of observation and inspection).

Thanks to the success of the Stockholm Accord, further agreements on conventional forces in Europe were negotiated (the Vienna Accord, the Conventional Forces in Europe Treaty (CFE), the Open Skies Treaty). All of these agreements have met with success, although Russia has warned it will not be in compliance with the CFE Treaty at the end of the three year reduction period in November 1995. (This subject is further explored in Chapter 14.)

In the Conference on Disarmament the Chemical Weapons Convention was successfully negotiated, but with less stringent on-site inspection requirements than first postulated. As it is yet to come into force, it is not possible to say how the verification provisions will be viewed in practice. In addition, the Biological Weapons Convention is undergoing a process whereby confidence-building measures and verification provisions are being worked out and will be integrated into the treaty in the next few years.

Current verification concerns

Nuclear nonproliferation

The mainstay of efforts to prevent the spread of nuclear weapons is the Nuclear Non-Proliferation Treaty (NPT). This treaty, negotiated between the years of 1965 and 1968, entered into force on 5 March 1970 and now has over 180 member states and was extended indefinitely at the NPT Review and Extension Conference in 1995. Adherence to the treaty is monitored by the International Atomic Energy Agency (IAEA) in Vienna through bilateral safeguards agreements between the Agency and each member state, but IAEA membership is not the same as NPT membership.

The NPT was severely undermined in the early 1990s by the discovery of the Iraqi nuclear weapon programme, the suspicion over the capabilities and intentions of North Korea and the long refusal of North Korea to fulfil its safeguards obligations, and the protracted dispute over the ownership of ex-Soviet weapons on Ukrainian soil. North Korea's agreement with the USA on the freezing of DPRK's nuclear weapons capability and the substitution of its current reactor programme with light water reactor technology has begun to ease the situation. However, the issue of challenge on-site inspections to undeclared sites is likely to cause problems in the future when, in approximately 5 years' time (1999–2000), these inspections are supposed to take place.

On the plus side, the destruction of South Africa's nuclear weapons demonstrated that while it may not be possible to 'disinvent' nuclear weapons, it is possible to verifiably dismantle them and the whole complex of weapons production facilities.

The Iraqi situation demonstrated deep flaws within the safeguards system—lack of resources within the IAEA; inspection criteria leading to numerous inspections of German, Japanese and Canadian installations and a only a handful of inspections of Iraqi facilities; and the failure of the international community to promote the use of special or challenge inspections. Challenge inspections (or in IAEA parlance, 'special inspections') are still of major concern. Apart from the enforced inspections in Iraq, the IAEA has only carried out 'special' inspections in Iran—at their request. Its attempts to do so in North Korea brought about the DPRK's notification of withdrawal from the NPT.

IAEA safeguards are designed to detect the loss of a 'significant quantity' of nuclear material from a safeguarded facility within a 'conversion time' (before the state can turn the diverted material into a nuclear weapon). The values for the 'significant quantity' and 'conversion time' were set by the Standing Advisory Group on Safeguards Implementation (SAGSI) in 1977. Today they look high. They are defined as follows:

Significant Quantity:	Plutonium	8kg
	HEU (highly-enriched uranium)	25kg
	LEU (low-enriched uranium)	75kg
	U-233	8kg
Conversion Time:	Plutonium (Pu)	7–10 days
	HEU	7–10 days
	Pure oxides/nitrates	1–3 weeks
	Spent fuel Pu	1–3 months
	LEU/natural U	12 months

There is an intrinsic uncertainty in determining the amounts of nuclear materials at plants. Material balance calculations will usually contain 'material unaccounted for' (MUF) even when there has been no diversion and, over a period of time, the MUF will exceed the significant quantity for diversion. However, the largest constraint on the effectiveness of IAEA safeguards is the combination of the safeguards budget and the focus of inspection effort. The frequency and intensity of inspections are currently determined by the amount of nuclear material put through each facility. As a result, until recently approximately 60 per cent of the IAEA's safeguards budget went into inspecting facilities in Canada, Japan and Germany. Now, thanks to Euratom, approximately 40 per cent of the inspection effort goes into inspecting Canada and Japan. This fact, coupled with the 1994 safeguards budget set at US$70 million and with a number of countries unable to make their contributions (e.g. Russia), the IAEA is severely stretched and unable to carry out its duties to the full.

Export Controls

In the early 1970s the Zangger Committee drew up an agreed 'trigger list' of equipment and materials which should only be exported under IAEA safeguards. In the mid-1970s the 'London Suppliers Club' (including France) extended the export controls to a wider range of technologies associated with reprocessing, enrichment and heavy water production plants ('sensitive technologies'). These Nuclear Suppliers' Guidelines also apply to re-exported items.

In 1992, the Nuclear Suppliers, Group, as it was then called, further strengthened export controls by agreeing on a common list of dual-use technologies and undertaking not to transfer nuclear facilities, equipment, components, material and technology to non nuclear-weapons states unless they accept full-scope IAEA safeguards. Both the Zangger list and the Nuclear Suppliers' Guidelines are codes of conducts rather than binding agreements. Industrially underdeveloped states strongly resent the guidelines, believing them to be devices to preserve the privileges of the industrially developed nations and to be in violation of Article IV of the NPT.

In addition, despite the strengthening of the guidelines and the increasing concern over the spread of nuclear weapons, companies in the UK exported sensitive technologies to Iraq. Although Iraq received a number of key technologies from the west, much of its nuclear weapons effort was based on old technologies which had been further developed by Iraqi scientists.

When considering the capabilities of states, it is important to remember that the basic technology for building nuclear weapons dates back to the 1940s. There are few states today which, if they so desired and if they had the resources in terms of cash and expertise, could not build a crude Hiroshima-style nuclear bomb. In doing so, such states may not need to import many items classed as 'sensitive technology'.

The INF and START (I and II) verification regimes

All three of these nuclear weapon reduction treaties have stringent verification regimes, the most comprehensive of which is that of the Intermediate-range Nuclear Forces treaty. This is because it is a 'zero option' treaty—all of the INF missiles have been destroyed and so any discovered now would be an unambiguous and serious violation of the treaty. The INF infrastructure has also been disbanded. 'Zero-option' treaties are always easier to verify. Because the START regimes leave the infrastructure intact and large numbers of weapons deployed, the degree of certainty in the verification regime is not as high as for the INF treaty.

The verification regimes for all the treaties (the START-II regime has yet to be implemented) include the following technologies and methods: on-site inspections; radiation detection techniques; production monitoring (including 'portal perimeter monitoring'); remote sensing by satellite; and witnessing destruction and reduction of weapons.

Weapons Conventions

Chemical Weapons Convention and verification

The Chemical Weapons Convention, signed in Paris in January 1993, has yet to enter into force. At the time of writing 35 states have ratified the treaty which needs 65 ratifications before it can enter into force and implement the verification provisions. (The UK has not yet ratified the treaty and the Queen's Speech on 15 November 1995 pledged to bring the Chemical Weapons Act on to the statute books in the 1995–96 parliamentary session.)

The verification regime of the Chemical Weapons Convention is extensive and intrusive: it includes inspections to commercial facilities. The implementation of the treaty, including the verification provisions, will be carried out by the Organization for the Prohibition of Chemical Weapons (OPCW) located in The Hague.

Each party to the treaty has to establish a National Authority to serve as the focal point for liaison with the OPCW and other parties. It is the National Authority to which the OPCW will turn first in order to resolve any non-compliance questions.

Biological Weapons Convention and verification

The 1972 Biological Weapons Convention has no verification provisions. However, since 1991, steps are being taken to include verification and confidence-building measures into the treaty. This is called the VEREX process.

In September 1994 the Special Conference to consider verification measures, which was held in Geneva, decided to set up an *ad hoc* group of experts. Proposals will be considered at the next Review Conference in 1996.

Anti-personnel mines

Land-mines or anti-personnel mines pose one of the most serious problems today. Owing to the efforts of the International Committee of the Red Cross, the issue of either banning anti-personnel mines or making them safe after a period of time is now on the international agenda. Both of these options pose severe verification challenges. If anti-personnel mines were to be banned completely then it would be possible to verify the cessation of production at declared facilities. It would, however, be much harder to stop or track the production at clandestine facilities. Verification of absence is obviously easier than verification of limits. If neutralisation devices were built into the mines so that they would become ineffective after a period of time, the installation could be verified at production. Again it would be much harder to stop clandestine production, but with the alternative of legal 'safer' mines, the market for illegally produced mines would be lessened.

The future

The key role for verification is increasingly one of confidence and security-building. If verification techniques and practices were included as one of the most significant engines in the machinery of security, then international security could have stronger foundations than today.

Nuclear nonproliferation

For nuclear nonproliferation, a new approach to verification is needed. Apart from political measures to strengthen the NPT, there are a number of technical steps that could be taken to increase the effectiveness of safeguards. In 1995, the IAEA initiated a series of steps that will strengthen the safeguards regime. The new programme (called 93+2 after the year the study was initiated, 1993, and the length of its duration) include the following steps: no-notice inspections; inspection of nuclear-related facilities; increased inspections in 'at-risk' states; and environmental monitoring. The full programme is not yet acceptable to many states, however, and could take many years to implement. The list below briefly summarises vital further improvements to the IAEA safeguards which are urgently needed to deter further nations from non-compliance.[6]

- A worldwide nuclear transparency system for reporting imports, exports and production of all nuclear materials and sensitive non-nuclear materials and equipment.
- A reduction by 50 per cent of the amounts of fissile material deemed to constitute a 'significant quantity'.
- Nuclear-weapons states accepting enhanced full-scope safeguards.
- Increasing the IAEA budget (but divorcing the increase from any increase in the promotion of civil nuclear energy budget) to enable the agency to carry out the necessary inspections.

One of the prerequisites for nuclear nonproliferation is to reduce the arsenals of declared and undeclared nuclear weapons states. There are two approaches to reductions in nuclear weapons: (i) the traditional approach of 'top down' reductions (that is, USA and Russia first, eventually bringing in China, France and the UK and hoping for cooperation from Israel, India, Pakistan, etc.); and (ii) the regional approach (that is, establishing regional nuclear weapons free zones through treaties such as the Treaty of Tlatelolco and the Treaty of Rarotonga). Both of these approaches require stringent verification. The first requires the type of verification regime set in place by the INF and START treaties, bringing other nuclear weapon states into the structure as and when necessary. The second needs a confidence-building approach as the states in the region build trust in each other through a sequence of agreements and confidence-building measures.

The regional approach could start with the two nuclear weapons free zones in existence (South Pacific and Latin America) and build from there. For example, in 1995 a treaty established an African Nuclear Free Zone. The regions of South Asia, the North Pacific/East Asia, the Middle East, North America, the CIS and Europe could all be involved in a process of confidence-building, transparency and where appropriate, nuclear weapons reductions in parallel with the 'top-down' reduction process.

Regional confidence-building and sub-state conflicts

The techniques developed to verify compliance with arms control agreements can also be applied to the building of confidence and trust as a means of conflict prevention. A mechanism for reducing tension between groups within states before it reaches the point of conflict is needed within the international security system. In the first place we need a mechanism for alerting the international community to tension which may escalate to violent conflict. Second we need a process of mediation for the hostile parties and third we need a set of tried and tested verification and confidence-building measures which could be appropriately chosen for individual situations. The application of verification and confidence-building measures to sub-national conflicts has received little attention until now and it is this new task for verification and confidence-building which provides us with the biggest challenge for the future.

The main difficulty in applying confidence-building measures to sub-state and trans-state conflicts lies in the fact that the situation is not one of state-to-state, but groups within states. Often, one of the groupings will be the government of the state or one of the groupings may inhabit a region across state boundaries. There may be many vested interests in not allowing a mediation and confidence-building process to begin. These include an unwillingness to share power, a fear of exposure, deeply held prejudices etc. There has, therefore, to be a procedure whereby a group or groups which feel under threat can approach the international community (the CSCE, the UN, etc.) directly and be accorded some status so that they may be recognised and heard internationally.

Although there are few state governments in a position to tolerate groups which directly oppose the state, or wish for more autonomy, being accorded international status—particularly if the groups have participated in violent action—it is possible for governments and sub-state groups to put aside such objections when a violent situation has become unbearable. During the process of mediation and negotiation, there are steps which can be taken to increase confidence in the intentions of the parties and to increase the likelihood of subsequent agreements succeeding. These range from building trust between local communities to providing data on levels of military equipment held by the state and by para-military organisations.

Civilian confidence-building measures

Building trust between, say, local communities or between, say, the state and a minority group through structured and agreed procedures is called civilian confidence-building. The role of verification and confidence-building in sub-state and trans-state conflicts, such as ethnic or religious conflicts, is a new idea.[7] Such measures could include:

- the setting up of youth organisations which include representation from all sections of the population;
- establishing an independent newspaper which is mandated to take the concerns and aspirations of all sections of the population into account and to help build bridges between minorities and majorities and which is monitored by an independent agency;
- establishing locally based committees, on which UN representatives also sit, to act as a forum for low-level complaints to lessen the risk for escalation into violent conflict;
- to set up, if appropriate, visits from communities in neighbouring states which have overlapping ethnic or religious communities to facilitate exchange of ideas, information and solutions, etc.

Military and para-military transparency measures

In the case of sub-national violent conflicts where there are national military or para-military or militia (as in the case of Northern Ireland for example), the military capabilities of all the groupings need to be known and monitored and that information made available to all parties. Independent observers could be allowed to observe the military capabilities of each party so that each side has more confidence in the numbers they are given. Most importantly, trained mediators should be enabled to set up meetings and exchanges between the parties so that differences and concerns over military and paramilitary capabilities and intentions can be aired and reconciled.

During a negotiation, it is unlikely that parties will wish to give highly detailed data on the location, configuration and command

and control of their military capabilities. However, once agreement has been reached a detailed data exchange and verification regime could be established through an independent organisation, such as the UN or the CSCE, and reductions, withdrawals, re-positioning and re-configuration could then be verified to everyone's satisfaction. Of course such a process cannot solve every problem and prevent every conflict. Some states and ethnic groups will refuse to be involved and some will lie and cheat on agreements reached. The earlier the process is started the higher the chance of success. For that reason the role of monitors and alerters requires urgent study. As in the case of state-to-state trust-building, such measures would not solve major problems by themselves and they are no panacea. However, they could help to reduce tensions and improve the climate for negotiations and long-term agreements. For example, in a long-running dispute, a number of confidence-building measures can relieve tension. Such measures can be reinforced if there is a degree of verification built in, for example through on-site inspections and aerial overflights. Tensions within negotiations are then reduced and parties can find that they reach agreement much more quickly—or they find that there is still disagreement, but it is no longer so critical. This process has been dubbed as '*agreeing when we can—negotiating when we can't*'. It takes a realisation that, although confidence-building measures cannot solve a problem, they do help to reduce tension and increase understanding and thereby facilitate creative discussion. The principle of 'agree when we can—negotiate when we can't could be the foundation for experimenting with a range of new civilian and military confidence-building measures and could set the scene for a more peaceful and prosperous twenty-first century.

Practical steps to increase security through verification

Verification is a process which establishes whether all parties are complying with their obligations under an agreement. These agreements can be international treaties on arms control or the environment, or agreements between different communities within a state. The success of any agreement depends on building an atmosphere of trust and trust can best be built and maintained when all sides are aware that cheating is likely to be detected. If carried out properly, with awareness of constraints, verification will always increase the security of the states and communities participating in the agreements. In this respect, verification is cost-effective. It builds confidence in the agreements and builds trust between the participants. The international community needs to make a concerted effort to promote multilateral arms reduction measures. This would need a decision to make issues such as nuclear nonproliferation, anti-personnel mines, biological weapons elimination, etc., the heart of foreign policy.

In the first place what is required is an increased commitment to research into new verification/confidence-building measures. In addition, the IAEA needs to do more than make minor modifications to the international safeguards regime. A fundamental rethink is needed. Regional confidence-and security-building measures could be promoted and they could begin to establish a framework for nuclear weapon free zones in the regions. From the regional confidence-building programmes, verification and confidence-building measures could be developed to also help solve sub-national conflicts.

If the international community were to approach multilateral disarmament and regional security through the verification and confidence-building route, it would build a more secure world built on knowledge and increasing trust rather than on threat and fear.

PART 6

RETHINKING THE UNTHINKABLE

16 REASSESSING THE NEED FOR BRITISH NUCLEAR WEAPONS

Michael Clarke

The next British government will inherit a nuclear weapons policy based on Cold War rationales. In the 1990s the Conservative government made some attempt to adapt to post-Cold War realities, phasing out land-based and air-delivered systems, so that by 1998 the British nuclear force will be concentrated on the Trident submarine force. The rationale for maintaining an independent British nuclear deterrent needs seriously to be re-examined, however, not least because the dynamics of nuclear proliferation have changed. The global situation suggests that nuclear proliferation will not get significantly worse for some time, and may now be on something of a plateau: but it could reach a stage in the next fifteen to twenty years where it might become considerably worse quite quickly, and the present world of eight or nine nuclear powers may suddenly become one of twenty or thirty. There is a breathing space, therefore, in which nuclear matters can be addressed against a reasonably stable background, and a window of opportunity in which considerable denuclearisation could take place.

The irony for the Labour Party is that at a time when so much has changed in the nuclear equation the Labour Party fights shy of nuclear matters, since it is now part of the folklore of the party that its anti-nuclear stance in the early 1980s lost it key electoral support. Many of the anti-nuclear arguments which were advanced in the late 1970s and early 1980s could now be regarded as vindicated by events, yet the Labour Party in opposition has been keen to deflect all debates on nuclear issues. There is also an irony in that the breathing space afforded to us during the next ten to twenty years allows any future government the option of taking no significant nuclear decisions, merely implementing the inherited policy. Decisions on a successor generation to the Trident nuclear force could be delayed into the early years of the next century, perhaps as late as 2005 to 2010, so there is no immediate pressure for the next government even to address nuclear issues beyond tactical questions of maintenance and auditing committed expenditure.

The reformed nuclear force structure

Britain's nuclear weapons are currently concentrated on the submarine force. Land-based nuclear missiles and artillery shells have been withdrawn, and the Royal Air Force will give up responsibility for dropping nuclear gravity bombs—the WE-177—by 1998.[1] *HMS Vanguard*, the first submarine in the Trident force, is already operational, *HMS Victorious*, the second, is on trials, and the other two boats in the force are now in the construction process. Firm orders have been placed for the fourth and final submarine, and cancellation of that boat will now be fairly expensive. By 1999, therefore, Britain's nuclear deterrent will rest entirely with the Trident submarine force. The government has said that each Trident submarine would not carry more than 96 warheads—though the boat and weapon design could accommodate 128—and 'may carry significantly fewer'.[2] Though the government will not say how many fewer than 96 could be deployed, its position is that 'the explosive power of the Trident systems as it is proposed to deploy them will be broadly the same as the Polaris system'.[3] The intention is that Trident will now cover both a strategic nuclear role and a sub-strategic role with the phasing out of the WE-177. That is to say, in the strategic role Trident is capable of inflicting unacceptable damage to the homeland of a potential aggressor against Britain; in the sub-strategic role, Trident warheads are regarded as sufficiently accurate to hit tactical targets and therefore could—at least in theory—be used in a war-fighting context.

For the future, therefore, Britain's nuclear weapons will be more coherently and simply structured than at any time in the past. Their military rationale is clear, even if most of the political arguments raise more questions than they answer, and the government claims that Britain's present nuclear force represents a considerable reduction from that of the Cold War; by 1998 it will possess 21 per cent fewer nuclear warheads than it did in the 1970s with a total explosive power some 59 per cent lower than at that time. This, it is said, demonstrates Britain's commitment to active nonproliferation.[4]

There are some important disagreements over these contentions, however. There are many who doubt the Trident system could play a sub-strategic role. There are doubts as to whether a sub-strategic nuclear role is feasible from a political point of view, since it is very difficult to believe that nuclear weapons would be used in anything short of a dire threat to the homeland itself. Even the protection of British troops deployed abroad would seem to most decision-makers too slight a reason to initiate a nuclear response. Then too, there are technical doubts over whether the Trident system would be sufficiently accurate and reliable to allow warheads to be used in an essentially tactical way. Certainly, it would appear that Trident could only be relied on in the sub-strategic role if it is also assumed that Britain continues to have some access to United States facilities for geodetic information, satellite guidance and pinpoint navigation over long distances. In an east/west context, it was relatively safe to assume that British missiles could hit Moscow, or major strategic targets (fixed sites, bases, command centres, etc.) in eastern Europe as part of the Single Integrated Operational Plan that would have come into effect in time of war. But in a more generalized political environment, in which British nuclear missiles may be targeted substrategically against an adversary as yet unknown, it is not clear that they could be directed at facilities further southwards, or even south of the Equator, with the same confidence in their accuracy.

There is also considerable debate over the contention that the Trident system will be configured to carry broadly the same general nuclear capability as Polaris. Certainly, the same explosive yield may be carried on Trident submarines, but it is distributed among a greater number of warheads—possibly three times as many—which are smaller, more accurate, and have a considerably greater range (12,000 miles as opposed to 4,600) than Polaris missiles. They therefore have the capacity to cover a wider range of targets. As the House of Commons Defence Committee candidly pointed out in 1994

> Trident's accuracy and sophistication does—and was always intended to—represent a significant enhancement of the United Kingdom's nuclear capability. We have invested a great deal of money to make it possible to attack more targets with greater effectiveness using nominally equivalent explosive power.'[5]

It is hardly surprising that the Trident submarine system represents a significant enhancement in capabilities over its predecessor of thirty years ago. It is a matter of political judgement, therefore, whether 'proliferation' should be regarded as increasing the amount of explosive power deployed by a country, or the number of targets its weapon systems can hit.

Rationales for British nuclear weapons

Since the demise of the Soviet Union, the rationale behind the possession and deployment of British nuclear forces has shifted somewhat. The independent British nuclear deterrent is now underpinned by five main lines of argument, the first of which is the bedrock rationale; the others are derivations of it.

1. The stated rationale for Britain being a nuclear power is the same as ever, namely that it is an ultimate guarantee of British security. This assumption was restated in February 1995 in more strident terms than ever before by Roger Freeman, the Minister for Defence Procurement, when he said 'One of the main tenets of our policy is that we should retain our nuclear deterrent while any other country in a position to threaten our security possesses a nuclear weapon, or the ability to construct a nuclear weapon'.[6] On this basis, Britain could retain its nuclear deterrent for ever since the *ability* to construct a nuclear weapon on the part of any country *in a position* to threaten British security is now a constant for the future. One of the reasons this statement is made in such strident terms may be the very vagueness of the potential threat. Whereas in the Cold War the nuclear threat to the British homeland was so specific it hardly needed any popular analysis—ministers merely had to remind the public what our deterrent was meant to deter—in the present environment the only clarity of threat lies in the theoretical potential of the technology. On Mr Freeman's analysis, Britain would be justified in retaining its nuclear weapons as a hedge against a US nuclear attack on Britain.

2. For historical reasons, Britain now possesses an independent nuclear deterrent, the acquisition costs of which—some £11 billion—have already been substantially met. Though there is dispute as to the extent of the running costs—which some analysts say could be as high as £30 billion over 25 to 30 years[7]—it is generally accepted that there would only be a limited peace dividend from cancellation of the Trident system at this stage.[8] There is an argument, therefore, that Britain's nuclear weapons represent an insurance policy on which the premiums have already effectively been paid, and in a

world in which nuclear weapons will have a lower salience, such a policy now represents a sensible hedge against an uncertain future. At a time when there are some encouraging developments around the world in nuclear nonproliferation, it is probably better to leave well alone and not raise major nuclear questions.

3. Closely allied to the second argument is the rationale that Britain represents a responsible holder of nuclear weapons. Douglas Hurd pointed out in January 1995 that the United Kingdom employs its nuclear capability in a 'thoroughly responsible way', which provides an example to other nuclear possessors and helps bolster a responsible nonproliferation regime based upon the present Non-Proliferation Treaty.

4. With Britain's economic performance falling behind most of its EU partners, there has been a tendency to re-emphasise the role of nuclear power as a symbol of influence and a source of respect among other states of the world. In the past, arguments about status were generally played down since they were in any case impossible to prove. Early statements in the 1950s that nuclear weapons guaranteed Britain a 'seat at the top table' were not often repeated, even if they were often remembered. They tended to attract derision in the case of a state in palpable retreat from an imperial role, and there seemed to be so much counter evidence. But in the post-Cold War environment the status argument has reappeared more overtly and with greater force, particularly in relation to Britain's position as a member of the United Nations Security Council. All five permanent members of the Security Council are nuclear powers and while nuclear status should not be regarded as a prerequisite for permanent membership, the renunciation of nuclear status might be regarded as weakening one's commitment to remain a member of the P5. Expressing views believed to be very close to those of the government, Sir Michael Quinlan, in an article in 1993, stated this point with some clarity: 'In a highly uncertain world the present would be a surprising time for the two permanent members of the UN Security Council [Britain and France], with no massive resource dividend in prospect, to decide to retire from nuclear activity and leave its burden entirely to the United States'.[9]

5. Finally, there is a seldom stated rationale for the present possession of nuclear weapons, which probably reveals most about the assumptions that British leaders make on the matter of proliferation.

Though policy-makers generally agree in private that it is almost impossible to conceive of the circumstances in which Trident would ever be used or its use threatened within its lifetime (1995 to around 2020); though they agree that Britain does not strictly need to operate an independent nuclear deterrent for its own sake; though they agree that Britain does not need to continue development on further nuclear weapon designs; and though they agree that continued nuclear weapons testing is not absolutely essential to the nuclear programme, the underlying assumption is that nuclear activities are vital, chiefly because they will keep alive the expertise to develop the next generation of weapons after Trident. In short, Trident is regarded as a post-Cold War bridge towards a new generation of weapons which—by the year 2020—*will* have a concrete rationale, since by then the international proliferation situation will be much less stable, and there will be no alternative but to enter into significant relationships of nuclear deterrence with a number of potential adversaries. To dispense with a complete weapon system during a period when we do not seem to need it will make it very difficult—perhaps even impossible—to resurrect it some time in the future, when we are almost certain to need it.

All of these arguments existed during the Cold War period, and were articulated from time to time with different degrees of emphasis. While the competition between NATO and the Warsaw Pact dominated Britain's security environment, however, it was merely sufficient to argue that the conventional superiority of the Warsaw Pact made NATO's nuclear weapons—of which Britain's force was a part—vital to compensate for this military disadvantage. With the end of Cold War rationales, arguments that were specific to the east/west balance in Europe became transformed into general deterrence arguments for which it is much more difficult to construct credible scenarios and which expose more clearly the underlying assumptions that are being made.

The twenty-year window of opportunity

In contrast to the atmosphere of the early 1980s, the nuclear picture is not becoming inexorably worse. In the short term, at least, there are some encouraging trends. No fewer than four states which inherited nuclear weapons have agreed to give them up, three of them as a unilateral act, one of them as part of a tripartite agreement.

Belarus, Kazakhstan, and Ukraine all inherited nuclear deployments from the Soviet Union upon its demise in 1991. The incoming South African government in April 1994 was also set to inherit a nuclear weapons programme. Belarus and Kazakhstan returned their nuclear weapons to Russia and began the decommissioning of others; the outgoing South African government dismantled the nuclear programme and the new government has forsworn the nuclear option; and after three years of protracted and difficult negotiations, Ukraine agreed in January 1994 under a tripartite deal with Russia and the United States, to remove its inherited nuclear weapons back to Russia for dismantling.

The Nuclear Non-Proliferation Treaty was extended indefinitely following the NPT Review Conference in April—May 1995, acquiring many new adherents in the process. Israel, India and Pakistan represent the only significant states still outside the NPT regime and the NPT Review Conference agreed new initiatives on Middle Eastern nuclear disarmament as part of the final package. In general, nuclear proliferation has not been as extensive or inexorable a process as was anticipated in the 1960s and 1970s. Many states approached the nuclear threshold, but remained on the non-nuclear side of it, and this situation is likely to endure for some time. For the next ten to twenty years, therefore, the number of declared and *de facto* nuclear states is not likely to increase appreciably. Instead, a number of countries will reach and remain below the nuclear threshold, not wanting to attract the opprobrium of crossing the line, but nevertheless having the capability to do so at relatively short notice should the need arise. By the year 2015 or 2020, therefore, the world may still have only eight or nine declared or *de facto* nuclear states.

On the other hand, regional instability could provoke a number of threshold nuclear states to go nuclear very quickly; perhaps almost simultaneously. As MccGwire points out, it would be prudent to anticipate that the world is more likely to revert to its longstanding pattern of proliferant behaviour than to continue along the present rather restrained track.[10] The history of the last 50 years is one of slow but steady nuclearisation, and though the present situation may be more relaxed, we should not assume that a world in which nuclear weapons have a generally low salience will prove to be the norm unless we make it so by more decisive political action. Space does not allow a fuller exposition of the critical argument between those who would be satisfied with a 'low salience' nuclear world and those who maintain that this is merely temporary and that a 'high salience' nuclear world is far more likely in the long run.[11] Account must also be taken of the fact that other weapons of mass destruction, particularly chemical and biological weapons, are becoming more generally available through civilian technologies to a wider range of countries. The US Department of Defense has estimated that some 24 states around the world have the capacity to produce significant amounts of offensive chemical and biological agents or have already done so. Other sources from the Russian foreign intelligence service and the US Office of Technology Assessment suggest a confirmed figure of around 15 states, with potential suspicions falling on several more.[12] Though there are potentially effective disarmament and arms control treaties governing chemical and biological weapons, the possibility remains that a 'breakout' of several nuclear threshold states could also provoke a similar breakout among states possessing chemical or biological potential who are not nuclear threshold states. On present trends, these possibilities are all in the pipeline, stored up for the moment, and perhaps emerging in around two decades time, when technological developments in the nuclear acquisition and ballistic missile technology fields intersect with regional tensions. Until then, the situation may be superficially stable.

We did not expect the world to give us such a breathing space. The speed with which the Cold War ended and the depth of dramatic political change in the east/west order has de-emphasised nuclear deterrence at a time when a number of nuclear proliferation issues had in any case to be faced. This period may therefore be regarded either as a window of opportunity in which significant measures of global denuclearisation could be adopted and a tight and verifiable nonproliferation regime constructed, or else merely a breathing space within the process of the inevitable nuclearisation of other parts of the world. Either way, this relatively brief period could end if and when a significant number of states reach the nuclear and missile technology thresholds in regional circumstances that are prone to instability. If not checked, the spread of ballistic missile technology around the world will be an important determinant of the end of any breathing space. If major adversaries can threaten each other with accurate ballistic missiles in regional contexts, in the absence of effective nuclear disarmament or really tight arms control measures, the temptations for other states to cross the nuclear threshold and break out of the NPT could become overwhelming.

Choices for a future government

The next government will therefore inherit a coherent nuclear force structure (which helps clarify the choices but which will offer few easy symbolic gestures for cosmetic denuclearisation), a window of opportunity in the international arena which will almost certainly never occur again, and a complex mix of rationales for Britain's nuclear forces built around a 'general deterrence' argument rather than a specific and identifiable military requirement. Of these three factors, the reading of the world situation is certainly the most critical. If a future generation of leaders believes that the present breathing space really is a window of opportunity and that significant denuclearisation in the world is possible, then the rationale behind present arguments will appear in a quite different light. If, on the other hand, leaders believe—as does the present government—that the ultimate outcome will be a world of many nuclear powers, in which nonproliferation efforts can slow but never reverse a trend towards nuclearisation, then it is difficult to quarrel with the logic presented in the five types of argument which bolster present nuclear forces.

For the purposes of analysis, Arguments 3 and 4 can be regarded as essentially peripheral. The image of Britain as a responsible decision-maker, or its status, either in general or on the United Nations Security Council, are arguments which bolster a central rationale but which are hardly central themselves. Certainly, an image of responsibility and the reputation of being an asset both to the nonproliferation regime and to the United Nations could be secured in other ways.

The essence of the case rests on Arguments 1, 2 and 5—that nuclear weapons provide Britain with an ultimate guarantee of its security, that they are a useful insurance policy in a 'low salience nuclear world', and that Britain will need to remain a nuclear power in the more likely event that the present period of 'low nuclear salience' gives way to one in which nuclear weapons acquire a much higher salience in the future. If a non-nuclear world is accepted as a realistic possibility (or even a world in which the possession of nuclear weapons is restricted, perhaps only to the United States, Russia and China, and monitored very closely by international bodies) then British security interests appear in a quite new light. On the first argument, it should be pointed out that many officials and ex-policy makers now accept—as Sir Michael Quinlan does—

that if Britain were *not* already a nuclear power then it would probably not now become one. In the post-Cold War world Britain's essential security can also be achieved by other means. It is not obvious that Germany, Italy, the Netherlands or Canada, for example, feel more vulnerable to nuclear missile attack than does Britain. Moreover, it is difficult for Britain to argue convincingly that we who are more territorially secure than at any time in our modern history find the possession of nuclear weapons indispensable to our future security, whereas other states in far more vulnerable situations than we should forswear them in the interests of world stability. Though Douglas Hurd is right to point out that Britain has not come under 'any coherent pressure to abandon its [nuclear] capability',[13] the fact remains that if threshold states at some time in the future decide to go nuclear, they will undoubtedly draw on the British (and French) examples of 'safe states' who feel that nuclear weapons are essential to them. In this respect, while the government may claim that Britain is a highly responsible possessor of nuclear weapons, Britain and France may come to seem in the long term to be the most irresponsible of proliferators, since they are non-superpower, middle-range states in a highly secure situation who refuse to relinquish a capability that analogous 'safe states'—most of them in the developed world—evidently regard as superfluous.

Far from being a relatively harmless and cheap insurance policy there is also the argument that such an insurance policy will prevent us from taking—or even realising that there exists—the favourable opportunities that the next few years offer to us; and worse, that the possession of nuclear weapons in British hands is more likely to usher in the period in which nuclear weapons take on again a high salience in world politics after this present breather. It is probably true that proliferation pressures around the world owe little directly to anything the British government does or does not do, but failure to grasp the opportunity of a genuine nuclear nonproliferation regime generates a pessimism regarding the long-term course of nuclear policy, and so becomes a self-fulfilling prophecy wherein the nuclearisation of the world seems inexorable.

Nor is it the case, under Argument 5, that any moves towards denuclearisation during the present breathing space are irrevocable. The deeply held argument that we must retain our nuclear expertise into the next era is, in fact, grossly overstated even on pessimistic assumptions. The present breathing space we have offers Britain a very safe international environment in which to denuclearise gradually. The 1995 Defence White Paper made it clear that long warning times could now be expected prior to the re-emergence of any strategic threat to the country.[14] There may well be time to reverse a policy of nuclear disarmament if it appeared to be counterproductive to Britain's security interests, and if that failed, there would be time to purchase nuclear weapons or vital nuclear components from elsewhere. It would be careless to assume that policies cannot be reversed; it is only that policy reversal can be politically difficult to accept at a sufficiently early stage and may in the event prove more expensive. These, however, may be regarded as relatively minor calculated risks well worth running in pursuit of a greater prize.

Optimistic assumptions regarding the next ten to twenty years therefore radically change the nuclear equation for any future government. Optimistic assumptions would seek to build upon the favourable developments in nuclear proliferation over the last few years, to construct a more complete and lasting nonproliferation regime. There is an interesting precedent in the Chemical Weapons Convention which is expected to enter into force during 1995 or early 1996. As a technology of mass destruction, the chemical weapons 'genie' has been out of the bottle for a great deal longer and has had a more mischievous existence than the nuclear 'genie'. Chemical weapons have been a recognised technology of warfare since 1915; they have been used on a number of occasions, and chemical weapons are far more inherent in civilian technologies than nuclear weapons ever could be. As genies go, the chemical genie is older, bigger and a more awkward customer than any of the other mass destruction genies we presently confront. Yet the world has now embarked upon a determined attempt to get this spectre effectively back into the bottle. A convention on the complete abolition of chemical agents as a weapon of war is about to come into force with tight and intrusive verification provisions. A complete spectrum of technology and an entire class of weaponry is about to be banned. Compared with this case, an optimistic approach to nuclear nonproliferation should not appear so impossibly idealistic. Of course, the Chemical Weapons Convention may not work with 100 per cent effectiveness. Nevertheless, there is a general assumption that an imperfect CWC is better than no CWC, and an imperfect CWC can only be built upon optimistic assumptions that it can be made to work, and that states will not pursue unilateral chemical weapons programmes as a hedge against the possibility that it will not be sufficiently effective.

For a future British government, therefore, the broad policy choices are relatively clear. If it takes an essentially pessimistic view of the possibilities of denuclearisation then it should do little to change the existing policy that it inherits. If one believes that we have no choice but to condemn ourselves eventually to live in a world of increasingly complex nuclear deterrence among a growing number of nuclear weapon states, then there will be little alternative to remaining a nuclear power. On such assumptions, the Conservative government's nuclear policy is now sensible, tidy and fairly prudent. If a minimum nuclear deterrent is to be maintained whilst others in the world also maintain nuclear forces, then this objective is most effectively achieved through the Trident missile system. Though the system has the capacity to represent far more than a minimum deterrent, the policy a new government would inherit involves commitments and declarations that it would only be deployed at minimum levels. One of the rationales for running a system that has spare capacity built into it is that over its 20–30 year life span the definition of what is 'minimum' may increase due to unforeseen circumstances. An incoming government that was as essentially pessimistic as the outgoing one, therefore, would simply declare that it needs to continue to maintain a minimum deterrent, though there could be some argument as to how the concept of a minimum will be interpreted. Such an approach might want to further reduce declared warhead numbers below the 96 per boat presently adopted as a ceiling; it might de-MIRV the missiles so that each submarine would carry no more than 16 single warhead missiles;[15] it might even reduce the number of firing tubes on Trident boats from 16 to 12 or 8 as a further gesture of restraint. In addition to adjustments in the Trident force, a future government might also support more enthusiastically than its predecessor the conclusion and verification of a Comprehensive Test Ban Treaty—expected to be signed in 1996; and might work to speed up progress towards a verifiable fissile material cut-off convention which would end the production and even the stockpiling of weapons-grade nuclear materials. All such initiatives would be welcome, reducing the profile of the independent deterrent during the (probably temporary) period of a 'low nuclear salience' world and would not preclude it from helping to maintain the present momentum toward greater nuclear arms control. None of these measures, however, would change the country's basic nuclear stance, which would remain a nuclear weapons state, enjoying the benefits of a breather before the nuclear proliferation situation takes a turn for the worse; doing nothing significant to avoid the nightmare which appears close to an eventual certainty when there are many nuclear powers in the world.

On the other hand, a government which began from the assumption that denuclearisation is possible, could exploit several opportunities over the next 5–10 years. A clear gesture in favour of the denuclearisation of the world would be simply to announce the abolition of the independent nuclear deterrent and begin its immediate decommissioning. This would not be technically difficult: it is much easier to decommission systems than to build them, and the loss of the sunk costs (just under £12 billion for the building of the whole system) would be set against the savings on running costs over the next 30 years. It is very unlikely that Trident submarines already in existence could be used for any other purpose and it would be irresponsible to sell them to another country. They would have to be regarded as a pure loss and broken up. In the present situation, it is unlikely that there would be a huge international backlash among NATO allies if such a unilateral announcement were made. During the early 1980s, United States' officials and politicians lent their weight to the Thatcher government's arguments by threatening dire consequences if Britain, under a Labour government, were to abandon its independent nuclear deterrent. In the present circumstances, it is difficult to believe that the US would react unfavourably or even that it would be greatly interested. Since 1985 the world has been punctuated with unilateral acts of nuclear arms reduction and even complete renunciation. A British declaration of renunciation would certainly be the most dramatic act to date, but hardly without precedent.

However, there could be considerable domestic backlash from such an announcement, and fierce resistance from the defence establishment and some sectors of British industry and society. Given the folklore of the Labour Party and the decade of neuralgia it has suffered on nuclear issues, a future Labour government would be unlikely to take such a dramatic step; certainly not in its early years, having been out of power for so long.

Nevertheless, a future government which wanted to seize the opportunities presented to it would have more choice than simply to continue Conservative policies with more accent on arms control, or abolish the independent nuclear deterrent altogether. It could make its nuclear policy on the clear assumption that the best outcome for British security would be global denuclearisation—not the maintenance of an independent deterrent at the highest level we can afford and decently get away with. This would imply putting the independent nuclear deterrent explicitly into future rounds of strategic arms control. Previous governments have frequently argued that the independent nuclear deterrent could not be subject to negotiations since it represents a minimum deterrent and nothing can be shaved from it without undermining its credibility. This is clearly no longer the case—if, indeed it ever was—since on the Conservative government's own admission the system has spare capacity which does not have to be used for a minimum deterrent function. It is quite credible, therefore, that the Trident system could be negotiated down from four boats to three or even two; or that four boats could be maintained with dramatically fewer warheads and each boat's capacity verified by intrusive international inspection. The essence of the approach would be that a future government would not be frightened of arms control and—if the conditions were favourable—would be prepared to give up the whole independent nuclear system as a major step towards the effective denuclearisation of the world. On this basis, a future government would support as enthusiastically as possible all the peripheral arms control initiatives such as the CTBT, a fissile material cut-off convention, and the urging of another round of strategic arms reduction talks. But the centrepiece would be an attempt to create for the nuclear world some counterpart to the Chemical Weapons Convention which will probably have just entered into force at the time when a new British government is taking office.

To make denuclearisation the main focus of British nuclear policy would therefore imply not just that a firm and serious commitment be made to the process of disarming national deterrents, but also that much greater effort is made to support and further develop the existing nuclear nonproliferation regime. The now indefinitely extended Non-Proliferation Treaty must be taken seriously; the International Atomic Energy Authority should be given far greater resources in order to conduct an adequate monitoring operation of nuclear installations around the world; and initiatives should be taken through the UN Security Council to increase the international pressure—and perhaps enforcement mechanisms—against states known to be in noncompliance with the NPT regime. The original NPT regime was always a bargain between the existing nuclear weapon states who agreed under Article VI to 'pursue negotiations in good faith on effective measures relating to cessation of the nuclear arms race at an early date and to nuclear disarmament', and those other states who agreed to forswear nuclear weapons indefinitely. This bargain can only be maintained and deepened if the non-nuclear weapon states are convinced that the existing nuclear weapon states are fully committed to their Article VI obligations.

If strategic arms reduction negotiations do not offer the prospect of effective denuclearisation and a deeper nuclear nonproliferation regime (in which, after all, Britain is only a medium player) then a complementary approach might be to work to institutionalise the nuclear deterrent within a European framework. Cooperative arrangements between the British and French independent deterrents are not now difficult to envisage and such arrangements could be institutionalised within a European nuclear planning group which could have the right of prior consultation—though perhaps not a veto, over the freedom of action of Britain and France.[16]

The ultimate objective of such an arrangement would be to wean London, and Paris, off the fixation with nuclear deterrence as essential to the security of the state. This, at least, would place the deterrent in a joint regional context where the forms of national control could be preserved as a harmless placebo, against the reality that it is impossible to conceive of British nuclear weapons being used in a genuinely unilateral context. If they ever become relevant, it would be in the context of *European* security; and it is infinitely more likely that they will not be relevant at all to European (or British) security problems. The strategic advantage of this arrangement would be to encourage more stability in Europe's relations with other actors in the world since it would embed potentially dangerous technologies within a multinational context. The disarmament advantages of the arrangement would be that it would allow the European nuclear powers a cheap and safe experiment with deterrence. If it were ever genuinely invoked then it would have more credibility in a European context than in a national one, and if it were never invoked then this institutional arrangement would help to establish the superfluity of nuclear forces for western countries, allowing the form to remain while the substance ebbs away for lack of a realistic purpose. Though this avenue would make far less of a contribution to the establishment of a tight nonproliferation regime and would not so clearly seize the opportunities for denuclearisation that now present themselves, it would, nevertheless, be a considerable improvement on the mere maintenance of an independent nuclear deterrent. It would open new opportunities for reductions and monitoring while providing a reassurance to those who cannot bring themselves to support a more decisive act of disarmament.

Conclusion

A future government will find it relatively difficult to criticise its immediate predecessor over the details of nuclear policy in the mid-1990s. Much has been changed since the mid-1980s and many of the most egregious nuclear follies have been, or are being, phased out. It will be faced with the immediate temptation simply to carry on with the existing policy since this offers the short-term prospect of a quiet life on the issue. This, however, can only logically and morally be pursued if the new government assumes that a much more dangerous world of many nuclear powers, in which Britain will have to play complex deterrence in a number of unspecified situations, is historically inevitable. If it does not begin with these assumptions then it must proceed on the basis that the best outcome for British security would be a denuclearised world—or a substantially de-nuclearised and heavily monitored world—in which Britain would not need, and would not possess, nuclear weapons. In this case, straight abolition of the independent nuclear deterrent would be the most desirable outcome, but for domestic reasons is probably impossible. More gradual moves in the same direction can be envisaged through strategic arms control or the Europeanisation of the British deterrent.

Whatever path the next government takes, however, it will be vital that it is prepared to think about the long-term implications of its actions. Nuclear history is at a potential turning point over the next decade, whereby the underlying situation could become dramatically worse or dramatically better. Successor generations will not easily forgive a failure of courage to think ahead beyond a single term of office.

APPENDIX

Statement on the Defence Estimates 1994[1]

The military tasks

DEFENCE ROLE ONE: to ensure the protection and security of the United Kingdom and our dependent territories even when there is no major external threat.

MT 1.1: Provision of an Effective Independent Strategic and Sub-strategic, Nuclear Capability
National nuclear capabilities, both strategic and sub-strategic, continue to underpin British defence strategy and provide the ultimate guarantee of our security. Maintenance of an effective independent strategic deterrent involves nuclear research, development, production and testing expertise and facilities; a minimum ballistic missile submarine (SSBN) force, providing assurance that at least one vessel can be at sea at all times, supported by secure, continuous real-time communications facilities covering the SSBN operating area: access to support and maintenance facilities for SSBNs, missiles and warheads; adequate conventional forces to safeguard deployment of the SSBN force; and conventional forces to safeguard at all times the physical security of nuclear assets as well as the command and control infrastructure.

MT 1.2: Provision of a Nuclear Accident Response Organisation
The Ministry of Defence would be the lead Government Department for the response to any incident or accident in the United Kingdom involving nuclear weapons, military nuclear materials or naval reactors. For this reason, the Department maintains specialist capabilities in support of the Nuclear Accident Response Organisation, designed to respond to such incidents or accidents.

MT 1.3: Provision of Military Support to the Machinery of Government in War
In transition to war or war the Government, at central or regional levels, would need to draw on military support to maintain the Machinery of Government in War. This requires provision of a military infrastructure, including communications and the maintenance of secure key points; military support for civil defence and emergency planning; and specialist support, including Explosive Ordnance Disposal teams.

MT 1.4: Provision of Military Aid to the Civil Power in the United Kingdom and Dependent Territories
Military Aid to the Civil Power (MACP) is provided in the United Kingdom and Dependent Territories for the direct maintenance or restoration of law and order in situations beyond the capacity of the civil power to resolve in any other way. The military role is to respond to a request for assistance, resolve the immediate problem and then return control to the civil power. MACP involves both specialist units—for example, bomb disposal teams—with the necessary specialist support and lift, and forces maintained for other tasks.

MT 1.5: Military Aid to the Civil Power in Northern Ireland
The armed forces continue to provide essential support to the Royal Ulster Constabulary (RUC) in Northern Ireland. This includes operations to deter and combat terrorist activity through the arrest of terrorists and the seizure of equipment and other resources; foot patrols to protect RUC officers carrying out normal police duties; vehicle check points; patrol bases at the border to discourage cross-border attacks; and specialist assistance, including helicopter support, bomb disposal and search teams.

MT 1.6: Provision of Military Assistance to Civil Ministries in the United Kingdom
Military Assistance to Civil Ministries is the use of military forces for non-military Government tasks, including assistance to maintain the essentials of life in the community or to undertake urgent work of national importance. Additionally, the Ministry of Defence routinely carries out a number of duties for other Government Departments on a repayment basis, in particular fishery protection, hydrographic tasks and assistance to HM Customs and Excise.

MT 1.7: Provision of Military Aid to the Civil Community
Military Aid to the Civil Community is the provision of Service personnel and equipment, both in emergencies and in routine situations, to assist the community at large.

MT 1.8: Provision of a Military Search and Rescue Service
The armed forces provide a continuous peacetime search and rescue (SAR) capability, with the priority task of rescuing Service personnel in the United Kingdom and surrounding seas. Where military SAR cover is not affected, SAR for the civil community is provided within the terms of agreements with other Government Departments. Under MTs 1.15, 1.18 and 1.20, SAR helicopter forces are also based in Cyprus, the Falkland Islands and Hong Kong.

MT 1.9: Maintenance of the Integrity of British Waters in Peacetime
The Government has an obligation to ensure the integrity of the United Kingdom's territorial waters and to protect British rights and activities in the surrounding seas. Military activities include sea and air surveillance of both surface vessels and submarines; maintenance of a presence in territorial waters and surrounding seas; and maintenance of the security of vital ports, anchorages and sea lanes, especially in time of rising tension.

MT 1.10: Maintenance of the Integrity of British Airspace in Peacetime
The integrity of British airspace in peacetime is maintained through a continuous Recognised Air Picture and air policing of the United Kingdom Air Defence Region.

MT 1.11: Military Intelligence and Surveillance
The armed forces assist the Government Communications Headquarters and other agencies in obtaining intelligence.

MT 1.12: Physical Security and Protection
The armed forces, the Ministry of Defence Police and Guard Service and, in some cases, civilian security services operate with the civil police and other agencies to protect Service personnel (both on and off duty) and their dependents against terrorist attack and to guard establishments, ships, aircraft, equipment and munitions against destruction or theft by hostile individuals or organisations.

MT 1.13: Provision of HMY Britannia and The Queen's Flight
The Government provides secure maritime and air travel, as required, for The Sovereign.

MT 1.14: State Ceremonial and Routine Public Duties
The Department provides military personnel for State ceremonial and routine public duties.

MT 1.15: The Security of Cyprus Sovereign Base Areas
The United Kingdom retains a substantial presence in the Sovereign Base Areas (SBAs) of Cyprus. This is centred on two resident infantry battalions and RAF Akrotiri, and provides communications facilities; an airhead for reinforcement, and evacuation when necessary; a Forward Mounting Base for operations in the Middle East and North Africa; military search and rescue; and training facilities for resident and non-resident forces. Other activities include assistance to the SBA administration, in particular maintenance of law and order in the SBAs using a mixture of military and civilian personnel.

MT 1.16: The Security of Gibraltar
The Government is responsible for the defence and the internal security of Gibraltar. The United Kingdom provides forces to deter and if necessary defend against aggression. We also provide a tri-Service Headquarters and facilities for the NATO commander for the Gibraltar area; operate, protect and maintain communications and surveillance facilities; provide a Forward Mounting Base and a Royal Air Force-manned airfield, which is also used by civilian airlines; assist in the training of the Gibraltar Regiment; and make available a destroyer or frigate at specified notice, as a guardship.

MT 1.17: Maintenance of a Base on Ascension Island
The Royal Air Force maintains a presence at Wideawake airfield on Ascension Island to support the Falkland Island airbridge and reinforcement plans and to act as a Forward Mounting Base for evacuation operations in sub-Saharan Africa.

MT 1.18: The Security of the Falklands and South Georgia
The United Kingdom continues to maintain a defensive capability in the South Atlantic. The garrison is tasked with maintaining the integrity of Mount Pleasant airfield as an airhead for reinforcement; defending other military installations in the Falkland Islands; providing for the safety of shipping and aircraft within the area; countering military action against South Georgia and the South Sandwich Islands; and providing 24-hour military search and rescue (SAR) cover and, when military cover is not affected, SAR for the civil community.

MT 1.19: Maintenance of a British Military Presence in the British Indian Ocean Territory (Diego Garcia)
A small naval party is based on Diego Garcia in the British Indian Ocean Territory to exercise the Government's sovereign rights. Its tasks include administrative functions on behalf of the Commissioner of the Territory.

MT 1.20: The Security of Hong Kong
The garrison's primary role is to demonstrate British sovereignty and to support the Hong Kong civil authorities in maintaining internal security and stability.

MT 1.21: Maintenance and Activation of Service Evacuation Plans
In cases where civil contingency plans prove insufficient to guarantee their safety, we maintain plans to evacuate British nationals at short notice from a number of countries.

MT 1.22: Reinforcement of the Cyprus SBAs, Gibraltar, Ascension Island, the Falklands and South Georgia, Diego Garcia, and Hong Kong
The Government is committed to the stability, security and defence of the Dependent Territories, including, if necessary, the restoration and maintenance of law and order. None of the resident forces for Military Tasks 1.15 to 1.20 are capable of meeting all possible contingencies and therefore could require reinforcement. Rehearsal of reinforcement plans contributes to deterrence, as well as maintaining expertise.

MT 1.23: Reinforcement of Other British Dependent Territories
The Dependent Territories which do not have resident British garrisons face no particular military risk. They could, however, face challenges to their internal security which might lead to a request for military support, drawing on our national intervention capability.

MT 1.24: Provision of Hydrographic Surveying and Geographic Services
Hydrographic surveying and geographic mapping and survey services are a defence responsibility because of the security aspects of providing hydrographic support for the strategic deterrent, anti-submarine warfare and mine countermeasures operations, the security aspects of a geographic support, and the need to preserve a uniformed field capability for operations and emergencies.

MT 1.25: Ice Patrol Ship
British sovereignty interests in the Antarctic are demonstrated by the annual deployment to the region, during the austral summer, of the Ice Patrol Ship HMS Endurance. The ice patrol task includes assistance to the British Antarctic Survey, hydrographic survey and meteorological work.

DEFENCE ROLE TWO: *to insure against a major external threat to the United Kingdom and our allies*

MT 2.1: NATO Nuclear Forces
NATO's Strategic Concept requires the maintenance of nuclear forces, including sub-strategic forces, based in Europe, but at much reduced levels. The United Kingdom has committed all its nuclear forces, both strategic and sub-strategic, to NATO.

MT 2.2: Maritime Immediate Reaction Forces
Maritime immediate reaction forces provide a small core of units held at the highest levels of readiness which can be deployed at very short notice in response to a crisis. They would form the nucleus around which the United Kingdom and its allies could deploy their rapid reaction or main defence forces.

MT 2.3: Land Immediate Reaction Forces
Land immediate reaction forces are designed to provide a multinational presence in potential areas of crisis. This role is currently undertaken by the Allied Command Europe Mobile Force (Land), which is capable of ACE-wide operations, acting independently or in conjunction with other forces.

MT 2.4: Air Immediate Reaction Forces
Air Immediate Reaction Forces are capable of deployment ACE-wide at short notice.

MT 2.5: Maritime Rapid Reaction Forces
Maritime rapid reaction forces will respond to a crisis which exceeds the capability of immediate reaction forces to deter or counter. Their high state of readiness and need to react to a wide variety of military situations calls for a pre-planned force mix and capability. Maritime rapid reaction forces could be formed into NATO Task Groups, NATO Task Forces or NATO Expanded Task Forces

depending upon the requirements of a particular crisis.

MT 2.6: Land Rapid Reaction Forces
The multinational ACE Rapid Reaction Corps (ARRC)—to which the majority of NATO nations are contributing—is the key land component of NATO's rapid reaction forces. It will provide the Alliance as a whole with the ability to respond quickly and effectively to any major threat to its security.

MT 2.7: Air Rapid Reaction Forces
Air reaction forces are required to provide a capability across the broad spectrum of mission types: offensive air support, tactical reconnaissance; and interdiction.

MT 2.8: Maritime Main Defence Forces
Maritime main defence forces are at lower readiness than those in the maritime reaction forces but could be used to supplement or reinforce these formations in an escalating crisis.

MT 2.9: Land Main Defence Forces
If not required by the ARRC, we will contribute land main defence forces capable of conducting operations under NATO command. With the creation of AFNORTHWEST, the landmass of the United Kingdom will fall within ACE for the first time, and its defence will encompass some of the activities previously defined as Military Home Defence.

MT 2.10: Air Main Defence Forces
The United Kingdom contributes to Airborne Early Warning operations, offensive, defensive and reconnaissance operations for SACEUR with association ground and air support and wartime search and rescue operations.

MT 2.11: Maritime Augmentation Forces
The United Kingdom provides a range of forces and capabilities to NATO's maritime augmentation forces. These will be held at the lowest readiness, and in peacetime will mainly comprise vessels in routine refit or maintenance which will not be available for short-notice deployment.

MT 2.12: Special Forces
The provision of highly trained Special Forces (SF) able to carry out specialised military tasks is of considerable value in NATO's high-level operational planning. SF provide a unique contribution a the

strategic and operational level, but they are also able to provide significant support to conventional formations; they can be committed in peace, crisis and war. The United Kingdom contributes Special Forces at theatre level to support reaction and main defence force deployments. These can carry out surveillance, reconnaissance, offensive action and military assistance operations independently or in conjunction with other units.

MT 2.13: Deployment and Logistic Support
All the forces covered by the previous Military Tasks need to be deployed and sustained. We achieve this using military and civil air transport aircraft and shipping as appropriate to the situation, linked with a complex network of supply agencies, both in the United Kingdom and overseas.

DEFENCE ROLE THREE: to contribute in promoting the United Kingdom's wider security interests through the maintenance of international peace and stability

MTs 3.1-3.5: Maintenance of a National Intervention Capability
A number of Military Tasks in Defence Roles One and Three require forces to be available on a contingency basis. For some, the United Kingdom is likely to have to act alone. For others, operations are likely to be based on a multinational response, probably under United Nations auspices. The armed forces need to be able to produce a graduated range of military options, from the employment of small teams of Special Forces to the mounting of an operation requiring the deployment of a division with maritime and air support, as circumstances demand. We have therefore identified for planning purposes an intervention capability from which appropriate contingency forces could be drawn as required.

MT 3.6: Humanitarian and Disaster Relief
When appropriate, and at the request of the Foreign and Commonwealth Office or Overseas Development Administration, British armed forces contribute to humanitarian and disaster relief operations, either on an individual basis or as part of a co-ordinated international effort.

MT 3.7: Provision of a Military Contribution to Operations Under International Auspices
The forces identified in Military Tasks 3.1 to 3.5 provide the ability

to contribute to operations under international auspices, in particular those of the United Nations, CSCE and WEU, and to NATO operations in support of United Nations or CSCE mandates.

MT 3.8: Operational Deployments Under Bilateral and Multilateral Agreements

The responsibility for the defence of Belize was assumed by the Government of Belize on 1 January 1994. The British military presence in Belize will in future take the form of a training operation for troops from the United Kingdom. The United Kingdom is committed to the stationing of Gurkha battalion in Brunei until 1998; full costs are met by the Sultan. In the run-up to the withdrawal of the Hong Kong garrison in 1997, the Brunei garrison provides the acclimatised reserve for Hong Kong. We also have jungle training facilities in Brunei. We maintain our commitment to the Five Power Defence Arrangements (FPDA), which provides for consultation in the event of a threat to the security of Malaysia or Singapore. The commitment involves provision of Headquarters staff for the Integrated Air Defence System Staffs and participation on an opportunity basis in FPDA exercises.

MT 3.9: Reinforcement of Brunei

In the event of an external threat to Brunei, and subject to the consultation stipulated in the exchange of notes between the two Governments, the British garrison may be deployed in support of the Royal Brunei Armed Forces; this may require reinforcement.

MT 3.10: Other Operational Deployments

The United Kingdom provides forces which contribute to the development of greater stability both within and beyond Europe. The Armilla Patrol provides reassurance and assistance to entitled merchant shipping in and around the Gulf area. It is also helping to enforce the remaining United Nations resolutions on trade with Iraq. The armed forces also provide assistance to combat the trade in drugs, where this can be done without detriment to the performance of other military tasks. Overseas visits, including ship visits, provide unique opportunities for contact with foreign armed forces, and thus have an important role to play in developing military links. They can also contribute to improving bilateral relations in other ways.

MT 3.11: Military Assistance and Combined Exercises

Military assistance takes place mainly in support of wide foreign policy aims; the defence objective is limited to promoting stability and military effectiveness in countries where we retain valuable facilities, including for transit and training, or where we have an obligation to assist in the event of a security threat. Such military training can make a significant contribution to regional stability by promoting military effectiveness and individual states' own perceptions of security.

MT 3.12: Arms Control, Disarmament and Confidence and Security-Building Measures

Under current treaties and agreements, the United Kingdom has an inescapable duty to host incoming inspections, and also has the right to make a certain number of outgoing inspections.

NOTES AND REFERENCES

1 Security Challenges

Michael Clarke

1. British armed forces have played a role in all these situations, though sometimes as little more than a diplomatic gesture, as in Somalia and Haiti.
2. There are vigorous debates over the motives to conquest of the Soviet Union during the Cold War. There is, however, no debate over the fact that if the Soviet Union found itself—for whatever reason—involved in a general war with NATO, then its only credible military option was to try to conquer the whole of western Europe to eject the USA from the continent and deny it the base from which to launch a counter-attack. The fact that the Soviet Union never showed the slightest political desire to achieve this does not alter the fact that it represented its only sensible military option in the event of a world war. For Moscow it was the lesser evil. This is what the Warsaw Pact was structured to attempt to do, even if many of its commanders doubted that it could be achieved.
3. Some have argued that mere revenge or simple fanaticism may provide a motive to visit mass destruction on the homeland. In logic this is true—as it is throughout domestic society where a small minority of psychopaths are always at large. This is not normally regarded, however, as an intolerable threat, or one which should lead society to change the natural patterns of behaviour of all its other members.
4. On ballistic missile threats see Martin Navias, *Going Ballistic: the Build-up of Missiles in the Middle East*, London, Brassey's, 1993.
5. For a general analysis of European potential conflict points see Hugh Miall, 'New Conflicts in Europe: Prevention and Resolution', *Current Decision Report 10*, Oxford Research Group, 1992. See also, by the same author, *The Peacemakers: Peaceful Settlement of Disputes Since 1945*, London, Macmillan, 1992.
6. A good general review of these five areas can be found in Stephen I. Griffiths, *Nationalism and Ethnic Conflict: Threats to European Security*, SIPRI Research Report 5, Oxford, Oxford University Press, 1993.

7. Samuel Huntington, 'The Clash of Civilisations', *Foreign Affairs*, 72(3), 1993, p.35.
8. See, for example, Misha Glenny, *The Fall of Yugoslavia*, New Edition, London, Penguin Books, 1993, pp.235–42.
9. See Graeme Herd, *et al.*, *The Coming Crisis is Estonia*, London Defence Studies 28, London, Centre for Defence Studies, 1995.
10. Jed Snyder, 'Russian Security Interests on the Southern Periphery', *Jane's Intelligence Review*, 6(12), December 1994, pp.548–51.
11. Miall, op. cit., p.35.
12. See John Morrison, 'A Non-Involvement Option for Britain', in Michael Clarke and Philip Sabin (eds), *British Defence Choices for the Twenty-first Century*, London, Brassey's, 1994.
13. In Germany, for example, there are so many different immigrant groups from other parts of Europe that almost any southern or eastern European conflict raises the possibility of inter-communal conflict within the German Republic.
14. Around half of all US and Japanese investment into the EU still comes to Britain, and Britain remains—by a long way—the most powerful overseas investor in the EU. See Michael Clarke, *British External Policy-making in the 1990s*, London, Macmillan for the RIIA, 1992, p.48.
15. *Statement on the Defence Estimates 1995: Stable Forces in a Strong Britain*, Cm2800, HMSO, 1995, p.9.
16. *Statement on the Defence Estimates 1992*, Cm1981, London, HMSO, 1992, p.9.
17. *Statement on the Defence Estimates 1995*, op. cit., p.9.

2 Roles, Missions and Resources

David Greenwood

1. *Statement on the Defence Estimates 1993, Defending Our Future*, Cm2770, London, HMSO, 1993.
2. Ministry of Defence, *Front Line First: The Defence Costs Study*, London, HMSO, 1994.
3. Her Majesty's Treasury, *Financial Statement and Budget Report, 1995/96*, London, HMSO, 1994 (hereafter *Red Book 94*).
4. Malcolm Rifkind *et al.*, *The Framework of UK Defence Policy: Key Speeches on Defence Policy by Malcolm Rifkind QC MP 1993–1995 with Contemporary Commentaries*, London, London Defence Studies, No.30–31, December 1995.
5. *NATO Review*, April 1995.
6. Memorandum on the United Kingdom Government's Approach to the Treatment of European Defence Issues at the 1996 Inter-Governmental Conference, March 1995.

7. The United Kingdom Defence Programme: The Way Forward, Cm8288, London, HMSO, 1981.
8. See D. Greenwood, 'Expenditure and Management', in P. Byrd (ed.), *British Defence Policy: Thatcher and Beyond*, London, Philip Allan, 1991, pp.36–66.
9. The ceiling was set at £2 billion, at 1964 prices, below which total spending was to be brought by 1969/70.
10. *Statement on the Defence Estimates 1991, Britain's Defence for the 90s*, Cm1559-I London, HMSO, 1991. (Cm1559-II is the companion volume to *Defence Statistics 1991*).
11. See the account in my contribution to M. Clarke and P. Sabin, *British Defence Choices for the Twenty-First Century*, London, Brassey's, 1993.
12. *Statement on the Defence Estimates 1992*, Cm1981, London, HMSO, 1992, pp.8–9.
13. Cm2770, cited at note 1 above.
14. Ibid., Table 3
15. Ibid., Tables 4 and 5.
16. See D. Greenwood, 'The Post-Cold War Politics of "Muddling Through"', *Enjeux Atlantiques*, July 1994, pp.42–8.
17. *Red Book 94*, cited at note 3 above.
18. *Red Book 94*, Tables 6.5 and 6A.3. See also the Public Finance Foundation's commentary *The Budget 94*, London, Public Finance Foundation, with Ernst and Young, 1994, p.27.
19. See D. Greenwood, 'Expenditure and Management', cited at note 8.
20. The table is a compressed version of that in the *Front Line First* report, Table 2, p.39.

3 Restructuring the British Army

Colin McInnes

1. On the ARRC see Colin McInnes, *The British Army and Nato's Rapid Reaction Corps*, London Defence Study 15, London, Brassey's, Centre for Defence Studies, 1993.
2. *Statement on the Defence Estimates 1994*, London, HMSO, 1994, p.57.
3. *Front Line First: The Defence Costs Study,* London, Ministry of Defence, 1994.
4. *Fifth Report from the Defence Committee Session 1993–4, Implementation of Lessons Learned from Operation Granby*, London, HMSO, 1994, p.xiii.
5. *Statement on the Defence Estimates 1994*, p.37.

4 Weapons Procurement and the Defence Industry

Ron Smith

1. *Statement on the Defence Estimates 1994*, HMSO, para 430.
2. For instance, *The Downey Report, Development Cost Estimating*, 1966, *The Rayner Report, Government Organisation for Defence Procurement and Civil Aerospace*, 1971, and *Learning from Experience*, G. Jordan, I. Lee and G. Cawsey 1988. All are published by HMSO.

5 Rethinking British Arms Export Policy

Susan Willett

1. For further details see D. Miller (forthcoming) 'Motives and the Meaning of Guidelines in Arms Export Policy: The UK and the Iran–Iraq War', in S. Willett and M. Navias (eds), *European Arms Trade,* New York, Nova Science Publishers.
2. These revelations were based on statements by the Permanent Secretary of the Overseas Development Administration, Sir Tim Lankester, that £234 million in overseas aid had been used as a 'sweetener' to secure £1 billion of defence sales to Malaysia, in contravention of the government's own regulations concerning the allocation of aid. Further revelations were made concerning the improper promotion of certain favoured British companies—GEC and BAe—in the Malaysian arms deal, and the provision of subsidised financing.
3. Specifically an inquiry is looking into the role of Royal Ordnance in helping the German company Heckler and Koch evade the arms embargo by supplying Serbian soldiers with small arms and ammunition, and the role of ICI, Royal Ordnance and Allivane in an illegal European cartel selling arms to Iran during the 1980s. See 'Customs Inquiry into Arms Trade: British Companies Alleged to have Breached Weapons Sanctions', Tim Kelsey and David Keys, *Independent*, 3 February 1994.
4. In 1989, following allegations in the *Observer* of bribes connected to these deals the National Audit Office (NAO), the government's accounting watchdog, set up an inquiry. In May 1992 Sir John Bourn, NAO's chief, handed his report to the Public Accounts Committee. The chair of the committee, Mr Sheldon, decided that the report was too sensitive to make public—or even to display to other members of the committee. At the time the report's suppression shocked many MPs, yet Sheldon justified this move on the grounds that its revelations might affect future sales with Saudi Arabia. According to one source, the allegations in the report referred to British Aerospace paying hundreds of millions of pounds in bribes to the Saudi Royal family and to British

middlemen to secure the deal. See 'Ask me no questions, and I'll tell you no lies', *The Economist*, 12 February 1994, p.25.

5. 'Exporting British Arms', *The Economist*, 12 February 1994, p.20.
6. Ibid.
7. *Independent*, 29 June 1988.
8. See the *Guardian*, 29 December 1993.
9. Campaign Against the Arms Trade, *Death on Delivery*, London, CAAT, 1989, p.19.
10. R. Smith, 'How Will Limitations Affect the Overall Economies of Exporter Nations?' in *International Control of the Arms Trade*, Oxford Research Group Current Decisions, Report No.8, April 1992, and Ben Jackson, *Goldrunners Gold: How the Public's Money Finances Arms Sales*, World Development Movement, 1995.
11. Arms Control and Disarmament Agency, *World Military Expenditures and Arms Transfers*, Washington DC, US Government Printing Office, 1995 and National Audit Office, *Ministry of Defence: Support for Defence Exports,* London, HMSO, 1995.
12. Jackson, op. cit., p.15.
13. It has been estimated that the research and development costs of major weapon systems account for roughly 30 per cent of their unit costs, the government being the main source of defence R&D funding. Export sales, particularly of major weapons platforms, rarely generate sufficient economies of scale to spread these R&D costs significantly. In these circumstances recipient countries are benefiting from state subsidised R&D efforts rather than the other way around, as is often claimed.
14. The National Audit Office revealed that DESO had net operating costs of £8.5 million in 1987.
15. The defence share of export credits has grown to about half of the total provided by the Export Credit Guarantee Department. In 1993/4 £1,973 billion was provided to cover arms sales. See Jackson, op. cit., p.21.
16. The Pergau dam scandal exposed the use of overseas aid to 'sweeten' arms transfers. The amount offered for the 'uneconomic' Pergau project represents more than 25 per cent of the total value of the original deal struck in 1988. However, since the MOU was signed, Malaysia has failed to follow through on the orders for the Tornadoes (preferring MiG-29s) or frigates. In value terms these items would have represented a significant part of the £1 billion order. This suggests that the aid provided to secure the deal now represents a much larger percentage of the total value of the deal, thus resulting in a much poorer rate of return to the UK economy. It would appear that the main beneficiaries will be companies such as Balfour Beatty, Trafalgar House and GEC-Marconi who are bidding for construction contracts for the dam, while the British taxpayer, already overburdened by tax increases, will be the ultimate loser. See Vivek Chaudhury and Simon Beavis, 'Inquiry Urged into Aid "sweeteners"' *Guardian*, 12 February 1994.

17. Smith, op. cit., p.12.
18. See for instance W. Hartung, *Conflicting Values, Diminishing Returns: The Hidden Costs of the Arms Trade*, New York, World Policy Institute, 1994, p.15.
19. Typically offsets which include licensed or joint production of a weapon system or component involve the transfer of technology which contributes to the process of conventional weapons proliferation.
20. Hartung, op. cit., p.15.
21. National Audit Office, op. cit., pp.8–9.
22. Lora Lumpe, *Sweet Deals and Low Politics: Offsets in the Arms Market*, FAS Public Interest Report, Jan/Feb., 1994, p.1.
23. National Audit Office, *The Risks to Value for Money in Defence Procurement*, paper presented to the 15th Conference of Commonwealth Auditor Generals, 1992, p.8.
24. Smith, op. cit.
25. National Audit Office, *Ministry of Defence: The Costs and Receipts Arising from the Gulf Conflict*, London, HMSO, 1992, p.1.
26. Joanna Spears, 'British and Conventional Arms Transfer Restraint', in M. Hoffman (ed.), *UK Arms Control in the 1990s*, Manchester, Manchester University Press, 1990, p.174.
27. Ibid., p.171.

6 Britain and Europe's Common Foreign and Security Policy

Trevor Taylor

1. Article J.1 of the Maastricht Treaty reads: 'The Union and its Member States shall define and implement a common foreign and security policy, governed by the provisions of this Title and covering all areas of foreign and security policy'.
2. Article J.4 of the Maastricht Treaty.
3. Western European Union, 'Platform on European Security Interests', The Hague, 27 October 1987, text supplied by WEU.
4. D. Dinan, *An Ever Closer Union: an Introduction to the European Community*, Boulder, Colorado, Lynne Rienner, 1994, p.467. Dinan offers a clear, concise account of the EPC machinery and history.
5. Lawrence Freedman, 'Great Powers, Vital Interests and Nuclear Weapons', *Survival*, Vol.36, No.4, Winter 1994–95, p.37.
6. Each issue of *Bulletin of the European Union* includes a section covering CFSP developments. The June 1994 issue, for example, addressed the common position adopted on the Serbia-Montenegro embargo and the statements made on behalf of the Presidency addressing former Yugoslavia, Angola, Ethiopia, Latvia, and Nigeria. The July–August

issue gave details of the statements on Nagorny-Karabakh, Burundi, Baltic states, former Yugoslavia, Gambia, Guatemala, Guinea-Bissau, Israel and Jordan, Kyrgyzstan, Lesotho, Nigeria, Nuclear nonproliferation, Rwanda, Tadjikistan, Occupied Territories, East Timor, Yemen and Zaire.

7. *Bulletin of the European Union*, No.7/8, 1994, pp.66–7, and No.6, 1994, p.83.
8. J. Lodge, 'The Transition to a CFSP', in J. Lodge (ed.), *The European Community and the Challenge of the Future*, Second Edition, London, Pinter, 1993, pp.227–8.
9. Lodge, ibid., p.231, notes Jacques Delors' distinction between common and single policy.
10. See, for example, Joseph J. Romm, *Defining National Security: The Non-Military Aspects*, New York, Council on Foreign Relations, 1993, p.6, quoting Richard Ullman, and p.85; Theodore Moran, *American Economic Policy and National Security*, New York, Council on Foreign Relations, 1993, is concerned with 'threats to America's ability to lead or influence others, in accordance with its own values, and to behave autonomously' (p.2).
11. High-level group of experts on the CFSP, First Report, *European Security Towards 2000: Ways and Means to Establish Genuine Credibility*, Brussels, December 1994.
12. These views were noted in 'Who are We in the World?', *Independent on Sunday*, 26 March 1995.
13. Lodge, op. cit., p.244.
14. 'Europe Needs a Defence Chief Says France', *The European*, 3–9 March 1995.
15. *European Security Towards 2000*, op. cit.
16. 'The Preliminary Conclusions on the Formulation of a Common European Defence Policy Examined by the WEU Council of Ministers in Noordwijk' also included the following points:
    'A common European defence policy will need to be formulated against the background of a thorough analysis of European security interests....'
    'The formulation of a common defence policy requires a detailed analysis of risks to European security.' See 'Europe Documents', No.1911, *Atlantic News*, No.88, Agence Europe, 22 November 1994.
17. Lodge, op. cit., p.230.
18. From WEU Council at Noordwijk, op. cit., paras 11–12.
19. 'EU Ministers Back Non-aggression Deal between NATO and Russia', *Financial Times*, 20 March 1995.
20. See, for instance, 'Some Financial Carrots but no Political Strategy', *The European*, 3–9 March 1995, on EU policy towards North Africa.
21. As suggested by the *European Security Towards 2000*, op. cit.
22. See the *European Security Towards 2000*, op. cit.

7 Sharing the Burden of European Defence

Malcolm Chalmers

1. International Institute of Strategic Studies (IISS), *Military Balance 1994–1995.*
2. Trevor Taylor, 'West European Security and Defence Cooperation', *International Affairs*, 70, 1, January 1994, p.11.
3. Stanley Sloan, *Burden-sharing in the post-Cold War World*, Congressional Research Service, 1993, quoted in Taylor, ibid.
4. IISS, op. cit.
5. James Adams and Andrew Stephen, 'So Long, Nice While it Lasted', *Sunday Times*, 12 March 1995.
6. 'European Security Policy Towards 2000: Ways and Means to Establish Genuine Credibility', *Report of High-level Group of Experts on the CFSP*, December 1994, p.3.
7. For example, see Malcolm Chalmers, *Biting the Bullet: European Defence Options for Britain*, IPPR, 1993, pp.10–17.
8. 'Le Jumelage', *Armed Forces Journal International*, December 1994, p.39.
9. 'France's Wandering Eye', *The Economist*, 26 November, 1994.
10. 'Wooing the WEU', *The Economist*, 4 March 1995
11. 'Towards a Euro-mix', *The Economist*, 11 March 1995.
12. 'Belgium, Netherlands Plan Joint Surface Fleet', *Jane's Defence Weekly*, 14 May 1994.
13. 'The Defence of Europe', *The Economist*, 25 February 1995.
14. *Statement on the Defence Estimates, 1994* London, HMSO, 1994, pp.14, 24.
15. Michael Shackleton, 'The Budget of the European Community: Structure and Process' in Juliet Lodge, *The European Community and the Challenge of the Future*, London, Pinter, 1993, pp.89–111.
16. Treasury, *Financial Statement and Budget Report 1995–6*, London, HMSO, 1994.
17. Secretary of State Malcolm Rifkind in evidence to House of Commons Defence Committee, *Statement on the Defence Estimates 1994*, *Sixth Report of the House of Commons Defence Committee*, London, June 1994, p.9.
18. Ibid., p.68.
19. 1993 figure. *SIPRI Yearbook 1994*, pp.400–1.
20. *Statement on the Defence Estimates 1992*, London, HMSO, 1992, p.32
21. J.R. Oneal, 'Budgetary Savings from Conscription and Burden-sharing in NATO', *Defence Economics*, 3, 1992, 113–25.
22. *NATO Review*, April 1994, p.35. NATO does not publish comparable figures for France.

23. All figures from *The Reality of Aid 1994*, Actionaid, 1994.
24. *Statement on the Defence Estimates 1991*, London, HMSO, 1991, p.26.
25. 'European Security Policy Towards 2000: Ways and Means to Establish Genuine Credibility', *Report of High-level Group of Experts on the CFSP*, December 1994, p.16.
26. During the period 1990/1 to 1995/6, defence spending is planned to fall by 16.6 per cent in real terms: an average annual reduction of 3.6 per cent in real terms. Treasury, *Financial Statement and Budget Report 1995–6*, London, HMSO.
27. For a more detailed discussion of the future of the army, see Chapter 3.
28. 'NATO's ARRC: shaping up for service', *Jane's World of Defence 1995*, p.22.
29. 'Britain's share of bill for Eurofighter nears £15 bn', *Sunday Times*, 12 March 1995; 'Progress on the Eurofighter 2000 Programme', *Third Report from the House of Commons Defence Committee*, May 1994, p.33.
30. Evidence from Nick Evans, Head of Resources and Programmes (Air), Ministry of Defence, to House of Commons Defence Committee, reported in 'Progress on the Eurofighter 2000 Programme', *Third Report from the House of Commons Defence Committee*, May 1994, p.7.
31. 'Ministers tightlipped as EF2000 takes bow', *Jane's Defence Weekly*, 14 May 1994, p.5.
32. Eric Grove, 'Keeping a World Class Navy', *Parliamentary Brief*, July 1993, p.33.

8 The European Industrial Defence Base

Philip Gummett

1. This chapter draws on work done in longstanding collaboration with Professor William Walker, University of Sussex, UK, Professor Judith Reppy, Cornell University, USA, and Dr Josephine Anne Stein, of PREST, University of Manchester, UK; colleagues in the CREDIT (Capacity for Research on European Defence and Industrial Technology) network; and on work contributed by other colleagues in the International Fighter Study of the Institute for Defense and Disarmament Studies, Cambridge, Mass.
2. A. Clark, Speech on 'Defence and the High Technology Industries', London, World Economic Forum, 4 September, 1991, Ministry of Defence News Release 106/91.
3. GRIP, *European Armaments Industry: Research, Technological*

Development and Conversion, Final Report for European Parliament/STOA, Brussels, European Parliament, Directorate General for Research, 1993.

4. Cabinet Office, *Forward Look of Government-funded Science, Engineering and Technology 1994*, London, HMSO, April 1994, para 3.10.
5. *Statement on the Defence Estimates 1994*, London, HMSO, April, 1994, para 505.
6. Ibid., para 437.
7. Ministry of Defence, *Front Line First: The Defence Costs Study*, London, HMSO, July, 1994.
8. Info-DGA, interview with Pierre Joxe, No.46, June 1992, Paris, ADDIM, p.5.
9. W. Dawkins, 'French N-research to Shed 4,000 [sic] jobs', *Financial Times*, 9 November 1991. (The 4,000 must be a misprint, and would make no sense given the scale of the military applications division of the CEA; the figure of 400 is given in the body of the article.
10. M. Porter, *The Competitive Advantage of Nations*, London, Macmillan, 1991 and R. Reich, 'Who is US?' *Harvard Business Review*, January, 1990.
11. I. Anthony, *et al.*, 'Arms Production and Arms Trade', *SIPRI Yearbook 1993*, Oxford, Oxford University Press, 1993.
12. Ibid., p.432.
13. W. Walker, P. Gummett, *Nationalism, Internationalism and the European Defence Market*, Chaillot Papers, No.9, Paris, WEU Institute for Security Studies, 1993.
14. W. Walker, P. Gummett, 'Britain and the European Armaments Market', *International Affairs*, Vol.65, 1989, pp.419–42, and also Walker W., Gummett, P., 1993, op. cit.
15. P. Gummett, J. Reppy, op. cit., and P. Gummett, 1991, op. cit.
16. R. Drifte, *Arms Production in Japan: The Military Application of Civilian Technology*, Boulder, Colorado, Westview, 1986.
17. M.W. Chinworth, *Inside Japan's Defense: Technology, Economics & Strategy*, London, Brassey's, 1992.
18. R. Samuels, *'Rich Nation Strong Army': National Security and the Technological Transformation of Japan*, Ithaca, NY, Cornell University Press, 1994.
19. J. Alic, *et al.*, *Beyond Spinoff: Military and Commercial Technologies in a Changing World*, Boston, Mass., Harvard Business School Press, 1992.
20. See R. Van Atta, 'US Dual Use Technology Policy', presentation by the Special Assistant for Dual-Use Technology Policy, Office of the Assistant Secretary of Defense, USA, to the Foundation for Science and Technology, London, 16 February 1994; and J. Gansler, 'Transforming the US Defence Industrial Base', *Survival*, Vol.35, No.4, 1993, pp.130–46.

21. J. Cooper, 'Transforming Russia's Defence Industrial Base', *Survival*, Vol.35, No.4, Winter 1993–94, pp.146–62.
22. A. Coghlan, 'Stony Ground for Britain's Ploughshares', *New Scientist*, 22 January, 1994, pp.12–13.
23. Defence Industries Council, *Government and the Defence Industry: Working Together*, London, DIC, 1993.
24. Commissariat général du plan, *L'avenir des industries liées à la defense*, produced by the Groupe de stratégie industrielle, Paris, La Documentation Française, 1993.
25. House of Lords, Select Committee on Science and Technology, *Defence Research Agency*, London, HMSO, 1994, HL Paper 24, session 1993–94. Also A. Coghlan, 1994, op. cit.
26. See A. Clark, 1991, op. cit.
27. See House of Lords, 1994, op. cit.
28. See Cabinet Office, 1994, op. cit.
29. *UK Defence Statistics 1993*, Table 1.5, gives as estimates for 1993–94: net intramural research £464 million; net extramural research £172 million; therefore, 73 per cent intramural.
30. Professor Alain Pompidou, House of Lords, 1994, memoranda.
31. *L'Armement: Revue de la Délégation Générale pour l'Armement*, Paris, No.29, October, 1991, p.3.
32. Commissariat général du plan, 1993, op. cit.
33. P. Lock, W. Voss, 'The German Case', in Gummett and Stein (forthcoming) op. cit.
34. B. Hagelin, 'The Swedish Case', in Gummett and Stein (forthcoming) op. cit.
35. For example, Brzoska and Lock, 1992, op. cit., and E. Sköns, 'Western Europe: Internationalization of the Arms Industry', in H. Wulf (ed.) *Arms Industry Limited*, Oxford, SIPRI, Oxford University Press, 1993.
36. High-Level Group of Experts on the CFSP, *First Report: European Security Policy Towards 2000: Ways and Means to Establish Genuine Credibility*, Brussels, Commission of the European Communities, DG1A, 1994.
37. Ibid., p.12.
38. I am grateful to Alain Deckers for supplying information on GARTEUR.
39. See L. Teisseire, 'Quelles institutions pour l'Europe de l'armement?' *L'Armement: Revue de la Délégation Générale pour l'Armement*, No.30, December, pp.32–8 and Borderas, rapporteur, Assembly of the Western European Union, *The European Armaments Agency—Reply to the Thirty-ninth Annual Report of the Council*, Paris, WEU, 1994, document 1219.
40. WEU Noordwijk Declaration, 14 November 1994.

9 Reassuring Central Europe

Jane M.O. Sharp

1. For a Hungarian view, see Attila Agh, *From Competition to Cooperation: The Europeanisation and Regionalisation of Central Europe*, Budapest, Hungarian Centre for Democracy Studies Foundation, Budapest Papers on Democratic Transition, No.98, 1994.
2. Adam Michnik, 'Toward Europe without Illusions', *Central European Economic Review*, May 1995, p.6.
3. CSCE membership expanded from 35 to 52 owing to the breakup of Czechoslovakia, Yugoslavia and the USSR.
4. Alex Pravda, 'Russia and European Security: The Delicate Balance', *NATO Review*, Vol.43, 3 May 1995, pp.19–24.
5. On Russian military aggression in the early 1990s see Renée de Nevers, 'Russia's Strategic Renovation', *Adelphi Paper*, No.289, IISS, July 1994; Fiona Hill and Pamela Jewett, *Back in the USSR: Russia's Intervention in the Internal Affairs of the Former Soviet Republics and the Implications for United States Policy Towards Russia*, Kennedy School of Government, Harvard University, January 1994.
6. Michael R. Lucas, 'The War in Chechnya and the OSCE Code of Conduct', *Helsinki Monitor*, Vol.6, No.2, 1995, pp.32–42.
7. Tom Buerkle, 'EU Leaders Clear Way for Accord with Russia', *International Herald Tribune*, 28 June 1995.
8. Vaclav Klaus speech cited by *Wall Street Journal* (Europe), 24 July 1995.
9. Christoph Bertram, 'NATO on Track for the 21st Century?', *Security Dialogue*, Vol.26, No.1, March 1995.
10. Audrey Choi, 'Clinton Urges Bonn to Lead on Unity, Stability in Europe', *Wall Street Journal* (Europe), 12 July 1994.
11. William Wallace, 'Germany as Europe's Leading Power', *The World Today* August—September 1995, pp.162–4.
12. Defence minister Volke Ruhe noted in October 1993, 'It is not in Germany's interest to remain a state on the eastern fringes of the western prosperity zone. We Germans are the first to feel the consequences of instability in the East', *IHT,* 8 October 1994.
13. Edward Mortimer, 'Disorderly Queue to join Western Clubs', *Financial Times*, 12 July 1995.
14. Virginia Marsh, 'Hungary Knocks Harder on EU Door', *Financial Times*, 19 July 1995.
15. Interviews in Warsaw.
16. Other states that have joined Association Agreements with the EU include Slovakia, Bulgaria, Romania, Estonia, Latvia, Lithuania, Cyprus, Malta and Slovenia.
17. Krakow Declaration, 6 October 1991.

18. Henry Kissinger, 'Be Realistic about Russia', *Washington Post*, 25 January 1994.
19. In a speech to the annual conference of the International Institute for Strategic Studies in September, 1993.
20. London interviews, 1994.
21. PfP Invitation, NATO Press Communique M-1 (94) 2, 10 January 1994.
22. Gunnar Lange, 'The PCC: A New Player in the Development of Relations between NATO and the Partner Nations', *NATO Review*, Vol.43, No.3, May 1995, pp.30–2.
23. Bruce Clarke, 'Meeting of NATO and Former Warsaw Pact Ministers becomes Trench Warfare', *Financial Times*, 11–12 June 1994.
24. Strobe Talbott, 'Why NATO Should Grow', *The New York Review of Books*, 10 August 1995.
25. Peter van Ham (ed.), *The Baltic States: Security and Defence after Independence*, Paris, WEU Institute for Security Studies, June 1995.
26. Greg McIvor, 'Strictly Neutral Sweden Shuts Down on European Defence Force', *Guardian*, 23 February 1995.
27. For more on early Soviet views on NATO see chapters by Jane M.O. Sharp and Hannes Adomeit in Neil Malcolm (ed.), *Russia and Europe: An End to Confrontation*, London, Pinter/RIIA, 1994.
28. Quoted in *Arms Control Reporter* and *SIPRI Yearbook 1994*.
29. For example, Article 5 of the *Joint Declaration of 22 States* signed at the CSCE summit in Paris, 19 November 1990.
30. Pavel Baev, 'Drifting away from Europe', *Transition*, Vol.1, No.2, 30 June 1995.
31. Andrei Kozyrev, 'Partnership for United Peaceful and Democratic Europe', *Frankfurter Rundschau*, 8 January 1994, reprinted in FBIS-SOV, 10 January 1994.
32. Boris Federov, 'The Cold War Will End Only When Russia Joins NATO', *Financial Times*, 20 September 1995.
33. Reported in the *Wall Street Journal* (Europe), 21 February 1995.
34. Charles Goldsmith, 'Russia Indicates Terms to Accept Larger NATO', *Wall Street Journal* (Europe), 13 March 1995; *Atlantic News*, No.2700, 8 March 1995, and *Atlantic News*, No.2702, 15 March 1995.
35. Andrew Marshall, 'Moscow Set Tough Terms for NATO Expansion Plan', *Independent,* 13 March 1995; Charles Goldsmith, op. cit.
36. Bruce Clarke, 'Signs from Russia Grow Stronger as Ministers Meet', *Financial Times*, 23 March 1995.
37. Pavel Baev, 'Drifting away from Europe', *Transition,* Vol.1, No.11.
38. I am indebted to conversations with Vladimir Baranovsky for this discussion.
39. NATO Press Office, *Study on NATO Enlargement*, Brussels, September 1995, p.16.
40. The NATO—Russia document title is *Area of Broad and Profound Dialogue and Cooperation Between Russia and NATO*.

10 Doing Business with the Former Soviet Union

Neil Malcolm

1. D. Yergin, T. Gustafson, *Russia 2010 and What it Means for the World*, London, Nicholas Brealey Publishing, 1994.
2. D. Allen, 'Can Containment Work Again?', *Survival*, Vol.37, No.1, Spring 1995, p.63.
3. Interview in *Vecherni Noviny* (Sofia), 30 January 1990, cited in *BBC Summary of World Broadcasts*, SU/0677, p.A1/2 (1 February 1990).
4. V. Petrovsky, 'Priorities in a Disarming World', *International Affairs* (Moscow), No.3, 1991, p.4. See also p.3: 'The History of International Relations Shows that Kant was Right.'
5. Interview on Radio Rossiya, 12 March 1992, cited in *Foreign Broadcast Information Service*, SOV-92-049 (12 March 1992); *New Times*, No.3, 1992, p.20.
6. 'A Transformed Russia in a New World', *International Affairs* (Moscow), No.4/5, 1991, p.86.
7. Local Russian military commanders, and some in Moscow, were suspected of fomenting these wars in the first place, partly to provide an excuse for intervention. See below.
8. Because of parliamentary objections, Yegor Gaidar had only been Acting Prime Minister. Gazprom was the state monopoly gas extracting, processing and exporting corporation.
9. Dmitrii Furman in *Moscow News*, No.37, 1993, p.7.
10. A. Pravda, 'The Public Politics of Foreign Policy' in N. Malcolm, A. Pravda *et al.* (forthcoming, 1996), *Internal Factors in Russian Foreign Policy*, Oxford, Oxford University Press.
11. For a case in favour of this means of combating Russian 'proto-imperialism', see Z. Brzezinski, 'The Premature Partnership', *Foreign Affairs*, Vol.73, No.2, March/April 1994, p.79.
12. BBC, *Summary of World Broadcasts*, SU/1626, p.B1, 2 March 1993.
13. 'Moskva gotvi svoe partn'orstvo za mir', *24 chasa*, 28 April 1995, p.12.
14. R. Asmus, R. Kugler, S. Larrabee, 'NATO Expansion: the Next Steps', *Survival*, Vol.37, No.1, Spring 1995, pp.20–5.
15. Andrei Zagorski, 'Russia and the CIS', in H. Miall (ed.), *Redefining Europe: New Patterns of Conflict and Co-operation*, London, Pinter/RIIA, 1994, pp.78–82.
16. Asmus, Kugler and Larrabee also support exploring this proposal. 'Nato Expansion: The Next Steps', *Survival*, Vol.37, No.1, p.23.
17. This option, and the potential security complications introduced by the EU—WEU—NATO relationship are discussed in M. Brown, 'The Flawed Logic of NATO Expansion', *Foreign Affairs*, May–June, 1995.
18. T. Taylor, 'Security for Europe', in H. Miall (ed.), op.cit., p.178.

19. James Mayall and Hugh Miall, 'Conclusion: Towards a Redefinition of European Order', in H. Miall (ed.), *Redefining Europe*, ibid., pp.265–6.
20. R. Keohane, 'Redefining Europe: Implications for International Relations', in H. Miall (ed.), *Redefining Europe*, ibid., pp.235–9.

11 Developing the Conflict Prevention Agenda

Andrew Cottey

1. This paper is a revised and expanded version of Chapter 3 of *The Pursuit of Peace: A Framework for International Action*, Bristol, Saferworld, September 1994. An earlier version of this paper was presented at the British International Studies Association Conference, University of York, 19–21 December 1994. The author would like to thank Saferworld and the many people who commented on earlier drafts of this paper. The development of this paper was greatly helped by discussions with United Nations staff, including officials from the UN Secretariat's Departments of Political Affairs, Humanitarian Affairs and Peacekeeping Operations. The contents are entirely the author's responsibility.
2. Text of a speech by the Prime Minister the Rt Hon John Major MP in Cape Town, South Africa, 20 September 1994, Press Office, Prime Minister's Office, London, p.8.
3. Boutros Boutros-Ghali, 'An Agenda for Peace: One Year Later', *Orbis*, Vol.37, No.3, Summer 1993, p.324.
4. Kumar Rupesinghe, 'Early Warnings: Some Conceptual Problems', *Bulletin of Peace Proposals*, Vol.20, No.2, 1989, p.184.
5. David Cox, 'Exploring "An Agenda for Peace": Issues Arising from the Report of the Secretary-General', *Aurora Papers 20*, Ottawa, Canadian Centre for Global Security, October 1993, p.9.
6. Boutros Boutros-Ghali, 'An Agenda for Peace: Preventive Diplomacy, Peacemaking and Peacekeeping', Report of the Secretary-General pursuant to the statement adopted by the Summit Meeting of the Security Council on 31 January 1992, New York, United Nations, 1992, pp.15–16.
7. Boutros-Ghali, 'An Agenda for Peace: One Year Later', p.324.
8. Ibid., p.325.
9. 'Preventive Diplomacy: A UN/NGO Partnership in the 1990s, Report of a Round Table on Preventive Diplomacy and the UN's Agenda for Peace', 28–30 January 1993, International Alert/United Nations University/National Institute for Research Advancement, p.5.
10. Articles 34, 35 and 99, Charter of the United Nations, Appendix B, Adam Roberts and Benedict Kingsbury (eds), *United Nations, Divided*

World: The UN's Roles in International Relations, Second Edition, Oxford, The Clarendon Press, 1993, p.509 and p.526.

11. Michael Kelly, 'Surrender of Blame', *The New Yorker*, 19 December 1994, pp.44–51.
12. Thomas G. Weiss, 'Intervention: Whither the United Nations?', *The Washington Quarterly*, Vol.17, No.1, Winter 1994, p.118.
13. Wilhelm Hoynck, 'CSCE works to Develop its Conflict Prevention Potential', *NATO Review*, Vol.42, No.2, April 1994, pp.16–22.
14. Thomas M. Franck and Georg Nolte, 'The Good Offices Function of the UN Secretary-General', Chapter 6 in Roberts and Kingsbury (eds), *United Nations, Divided World*, op. cit., pp.143–82; Kjell Skjelsbaek, 'The UN Secretary-General and the Mediation of International Disputes', *Journal of Peace Research*, Vol.28, No.1, 1991, pp.99–115; and *Preventive Diplomacy: A UN/NGO Partnership in the 1990s*, op. cit., p.5.
15. Boutros Boutros-Ghali, Secretary-General of the United Nations, 'Report on the Work of the Organisation from the Forty-seventh to the Forty-eighth Session of the General Assembly', New York, United Nations, September 1993, p.97.
16. 'Supplement to "An Agenda for Peace": Position Paper of the Secretary-General on the Occasion of the Fiftieth Anniversary of the United Nations', A/50/60 S/1995/1, 3 January 1995, paras 30–2.
17. Renée de Nevers, 'Democratisation and Ethnic Conflict', *Survival*, Vol.35, No.2, Summer 1993, p.39.
18. Tom J. Farer and Felice Gaer, 'The UN and Human Rights: At the End of the Beginning', Chapter 8 in Roberts and Kingsbury (eds), *United Nations, Divided World*, pp.240–96.
19. *Preventive Diplomacy: A UN/NGO Partnership in the 1990s*, p.6.
20. De Nevers, 'Democratisation and Ethnic Conflict', pp.33–4. See also Minority Rights Group (ed.), *Minorities and Autonomy in Western Europe*, London, Minority Rights Group, 1991.
21. Adam Roberts, 'The United Nations and International Security', *Survival*, Vol.35, No.2, Summer 1993, p.11; and Kamal S. Shehadi, 'Ethnic Self-Determination and the Break-up of States', *Adelphi Paper 283*, London, Brassey's for the IISS, December 1993, pp.75–80.
22. Ivo Daalder, 'The Future of Arms Control', *Survival*, Vol.34, No.1, Spring 1992, p.62; and Itshak Lederman, *The Arab-Israeli Experience in Verification and Its Relevance to Conventional Arms Control in Europe,* Occasional Paper 2, Centre for International Security Studies at Maryland, College Park, MD, School of Public Affairs, University of Maryland, 1989.
23. John Hawes, *Arms Control: A New Style for a New Agenda*, CISSM Papers 2, Centre for International Security Studies at Maryland, College Park, MD, School of Public Affairs, University of Maryland, 1993, pp.14–15.

24. Conclusion of the Presidency of the European Council held in Lisbon on 26–27 June 1992, EPC Press Release, 27 June 1992; and 'Principles Governing Conventional Arms Transfers', *CSCE Forum on Security Cooperation Journal*, No.49, December 1993.
25. Saferworld, *Arms and Dual Use Export Controls: Priorities for the EU*, Bristol, Saferworld, June 1994.
26. Jenonne Walker, 'International Mediation of Ethnic Conflicts', *Survival*, Vol.35, No.1, Spring 1993, p.105.
27. Franck and Nolte, 'The Good Offices Function of the UN Secretary-General', p.178.
28. Boutros-Ghali, *An Agenda for Peace*, pp.16–18.
29. Boutros-Ghali, 'Report on the Work of the Organisation from the Forty-seventh to the Forty-eighth Session of the General Assembly', pp.99–100.
30. The operational preventive functions of peacekeeping forces are discussed in the British Army's *Field Manual Wider Peacekeeping*, Third Draft, Chapter 3, 'Operational Tasks', pp.3.1–3.4.
31. Roberts, 'The United Nations and International Security', pp.21–2; and Gerald B. Helman and Steven R. Ratner, 'Saving Failed States', *Foreign Policy*, No.89, Winter 1992–3, pp.3–20.
32. Weiss, 'Intervention: Whither the United Nations?', pp.121–2.
33. Patricia Feeney, 'Fair Shares—for the Rich', *Guardian*, 11 March 1995.

12 New Coalitions for Peacemaking

Hugh Beach

1. John Chipman 'Managing the Politics of Parochialism', *Survival*, Vol.35, No.1, Spring 1993.
2. Marrack Goulding, 'The Evolution of United Nations Peacekeeping , *International Affairs*, Vol.69, No.3, July 1993, p.455.
3. For the mandate of UNPROFOR see Susan Woodward, 'United Nations Security Council Resolutions and Presidential Statements on Yugoslavia September 1991–January 1995', Appendix to *Balkan Tragedy, Chaos and Dissolution After the Cold War*, Washington DC, The Brookings Institution 1995, pp.401–24.
4. *International Herald Tribune*, 14 December 1992.
5. Ibid., December 1992.
6. Ibid., 27 June 1994.
7. Ibid., 30 September 1994.
8. See Renée de Nevers 'Russia s Strategic Renovation' *Adelphi Paper*, No.289, IISS, July 1994.

13 Conventional Arms Control Regimes for Post-Cold War Europe

Jane M.O. Sharp

1. 'NATO Arms Exports to Turkey and Greece: Inconsistencies Revealed', *Basic Notes*, London, July 1995.
2. Yeltsin cited in *Atlantic News*, No.2716, 12 May 1995, p.2.
3. For details of Russian attempts to evade CFE treaty limits see Jane M.O. Sharp in *The CFE Treaty: History, Analysis and Evaluation*, Oxford, SIPRI, Oxford University Press, 1996, forthcoming.
4. Major cited by R. James Woolsey, 'Russia is Playing the Great Game with CFE', *Wall Street Journal* (Europe), 5–6 May 1995.
5. 'Treaty Tinkering', *The Economist*, 4 November 1994; and Fred Hiatt, 'Russia Resents West's Hawkishness', *Washington Post*, 6 December 1994.
6. Poland, Hungary and Czechoslovakia issued unilateral statements at treaty signature in November 1990 making clear they signed as individual states.
7. Bruce Clark, 'Top Russian Soldier Scorns NATO', *Financial Times*, 18–19 March 1995.
8. Lukin interview in *Nezavisimaya Gazeta*, 14 March 1995, cited in OMRI, 15 March 1995.
9. Russian troops already patrol Georgia's border with Turkey, Kyrgystan's border with China, Turkmenistan's border with Iran. See Claudia Rosett, 'New States are Finding Russian Embrace Hard to Resist', *The Wall Street Journal* (Europe), 2 March 1995.
10. Kinkel cited by Reuters, 1 March 1995.
11. *Arms Control Reporter*, 1993, p.407, B.495.
12. *Arms Control Reporter*, 1994, p.407, B.506.
13. Interviews with Hungarian officials in March 1995.
14. Dimitur Mitkov, 'Kontinent Sofia' cited in the *Arms Control Reporter*, p.407, B.499.
15. Remarks made by the Romanian Ambassador in London, July 1995.
16. Statement by Lt. General G.N. Gurin to the JCG, 24 January 1995.
17. *Arms Control Reporter*, 1994, p.407, B.506.
18. On Norwegian concerns that CFE does not adequately address threats to the Nordic region, see Marco Smedberg, Robert Dalso and Hans Zettermark, 'War in the North within the Limits of the CFE Treaty', a paper presented at the seminar on *Arms Control and Nordic Security* at the Swedish Institute of International Affairs, 15–16 February 1993.
19. Ambassador Ole Peter Kolby, 'Statement to the JCG', 9 November 1993.
20. See for example, 'Ankara Reacts to Russia's Initiative on Arms Reduction', *Hurriyet*, 17 June 1993, English Translation in FBIS-WEU-93-123.

21. *Atlantic News*, No.2685, 13 January 1995 and interviews with British officials.
22. Interviews with German officials, March 1995.
23. Klaus Kinke, op. cit. (n.10).
24. James Adams, 'Russia Repudiates European Treaty on Force Reductions', *The Sunday Times*, 3 October 1994.
25. Interview material, see also *Jane's Defence Weekly*, 'NATO Acts on CFE Complaint', 7 October 1995, p.3; and 'Conventional Forces in Europe: Tanks and Flanks', *The Economist*, 18 November 1995, pp.52–5.

14 Priorities for Nuclear, Chemical and Biological Arms Control

Stephen Pullinger

1. 'The 1993 Chemical Weapon Convention', *ISIS Briefing*, No.32, January 1993.
2. Graham Pearson, 'Verification of the Biological Weapons Convention', *ISIS Briefing*, No.43, June 1994.
3. Notes on Security and Arms Control, Foreign and Commonwealth Office, No.12, January 1995.
4. *Statement on the Defence Estimates 1995*, London, HMSO, p.38.
5. Secretary of State Malcolm Rifkind, Hansard, 2 May 1995, c.157.
6. Foreign Office Minister, Douglas Hogg, Hansard, 14 January 1992, c.902.
7. Foreign Secretary, Douglas Hurd, evidence to the Foreign Affairs Select Committee, 18 January 1995, HC34-vi, Q.494, p.235.
8. Ibid., Q.498, p.236.
9. See 'The Size of Britain's Nuclear Forces', *Special ISIS Briefing Note*, April 1995.
10. Progress of the Trident Programme, Second Report of the Defence Committee, Session 1993–94, HC 297, Q.1177.
11. Ibid., para 26, p.xiv.
12. Ibid., para 6, p.vii.
13. Defence Minister, Nicholas Soames, Written Parliamentary Answer, Hansard, 4 April 1995.
14. G. Berdennikov, addressing the plenary session of the Conference on Disarmament, 23 February 1995.
15. UK Position paper, NPT/Conf. 1995/24, 21 April 1995.

15 Confidence Building through Verification and Transparency

Patricia M. Lewis

1. *Verification in all its Aspects: Study on the Role of the United Nations in the Field of Verification*, United Nations document, A/45/372, 28 August 1990.
2. Michael Herman, 'Intelligence and Arms Control Verification', *Verification Report 1991*, J.B. Poole (ed.), London, New York, VERTIC/The Apex Press, 1991.
3. Article 10 of the US draft of the CWC, Conference on Disarmament Document CD/500, Geneva, 1984.
4. Sidney Graybeal, US Commissioner to the US—USSR Standing Consultative Commission, 1973–77, quoted in Richard Scribner, Theodore J. Ralston and William D. Metz, *The Verification Challenge: Promise and Prospect of Strategic Nuclear Arms Control Verification*, Boston, Birkhäuser, 1985, p.21.
5. See for example, Gordon M. Burck, *The Chemical Weapons Convention Negotiations, Verification Report 1992*, J.B. Poole and R. Guthrie (eds), London, VERTIC, 1992, pp.126–8.
6. Owen Greene, 'Verifying the Non-proliferation Treaty: Challenges for the 1990s', VERTIC Report November 1992.
7. Dennis Sammut, *The CSCE and Russian Peacekeeping, Verification 1995: Arms Control, Peacekeeping and the Environment*, J.B. Poole and R. Guthrie (eds), VERTIC/Westview Press, 1995, p.291.

16 Reassessing the Need for British Nuclear Weapons

Michael Clarke

1. *Statement on the Defence Estimates 1995, Stable Forces in a Strong Britain*, Cm2800, London, HMSO, 1995, p.38.
2. Malcolm Rifkind, 'The Role of Nuclear Weapons in UK Defence Policy', *Brassey's Defence Yearbook 1994*, London, Brassey's, 1994, p.30.
3. House of Commons Foreign Affairs Committee, Second Report, 1994–95, *UK Policy on Weapons Proliferation and Arms Control in the Post-Cold War Era*, Vol.2, HC 34-II, p.40, Q.77.
4. House of Commons *Debates*, 28 March 1995, Vol.257, Col. 817.
5. House of Commons Defence Committee, *Second Report*, 1993–94, HC 297, para 26.
6. House of Commons, *Debates*, 16 February 1995, Col. 1146.
7. Greenpeace, *The True Cost of Trident*, London, 1992, p.1.

8. £30 billion over 30 years—if the running costs turned out to be that high—certainly represents a great deal of money, but an average saving of £1 billion per year would not, in itself, do much to restructure a defence budget currently standing at over £22 billion per year.
9. Michael Quinlan, 'The Future of Nuclear Weapons: Policy for Western Possessors', *International Affairs*, 69(3), 1993, p.493.
10. Michael MccGwire, 'The Possibility of a Non-Nuclear World', *Brassey's Defence Yearbook 1995*, London, Brassey's for the Centre for Defence Studies, 1995, p.351.
11. The best sources for these debates are the articles by Michael Quinlan (note 9) and Michael MccGwire, 'Is There a Future for Nuclear Weapons?', *International Affairs*, 70(2), 1994. See also MccGwire's arguments in 'Eliminate or Marginalize? Nuclear Weapons in US Foreign Policy', *The Brookings Review*, Spring 1995.
12. See Thomas Stock and Anna De Geer, 'Chemical Weapon Developments', *SIPRI Yearbook 1994*, Oxford, SIPRI, Oxford University Press, 1994, pp.315–16.
13. House of Commons Foreign Affairs Committee, Vol.II, op. cit., p.234, Q. 490.
14. *Statement on the Defence Estimates 1995*, op. cit., p.19.
15. A MIRV is a multiple warhead, carried on a single missile. Trident missiles are capable of carrying up to 12 MIRV warheads, though British Trident D5 missiles will be loaded for a maximum of 6 warheads.
16. See Michael Clarke, 'British and French Nuclear Forces After the Cold War', *Arms Control*, 14(1), 1993, pp.139–42.

Appendix

1. *Statement on the Defence Estimates 1994*, London, HMSO, 1994.

INDEX